Water-Soluble Polymer Applications in Foods

In memory of my late father, Eliezer, survivor of the holocaust, believer in Zionism and the foundation of the state of Israel, a hard-worker, warm family man, and book-lover; a sensitive and loving person who taught me that the reason and destiny of existence is study and knowledge.

Water-Soluble Polymer Applications in Foods

A. Nussinovitch

Blackwell
Science

Editorial Offices:
9600 Garsington Road, Oxford OX4 2DQ, UK
Tel: +44 (0)1865 776868
Blackwell Publishing, Inc., 350 Main Street, Malden, MA 02148–5018, USA
Tel: +1 781 388 8250
Iowa State Press, a Blackwell Publishing Company, 2121 State Avenue, Ames, Iowa 50014-8300, USA
Tel: +1 515 292 0140
Blackwell Publishing Asia Pty Ltd, 550 Swanston Street, Carlton South, Victoria 3053, Australia
Tel: +61 (0)3 9347 0300
Blackwell Wissenschafts Verlag, Kurfürstendamm 57, 10707 Berlin, Germany
Tel: +49 (0)30 32 79 060

First published 2003 by Blackwell Science Ltd

Library of Congress
Cataloging-in-Publication Data
is available

ISBN 0-632-05429-8

A catalogue record for this title
is available from the British Library

Set in 10/13 pt Times
by Integra Software Services Pvt. Ltd, Pondicherry, India
Printed and bound in Great Britain by MPG Books Ltd, Bodmin, Cornwall

For further information on Blackwell Publishing, visit our website:
www.blackwellpublishing.com

Contents

Preface

Water-soluble polymers are used to a major extent in the food and other industries. In general, they can be utilized as thickening and gelling agents, for the stabilization of emulsions, for syneresis control, as suspending agents, for coating and binding, among many others. The manufacturer selects hydrocolloids to achieve specifically requested targets. There are many books dealing with traditional food and other applications of hydrocolloids, or describing various interactions of the individual gums and their mixtures. The possibilities of using hydrocolloids are, in fact, unlimited, dependent only upon the user's imagination and inventiveness. In the last few years, many less conventional applications of hydrocolloids have been recorded in the literature. The purpose of this book is to present these ideas in detail, to provide the reader with a comprehensive theoretical and practical approach to these topics, as well as to provide tools for research and development, and to serve multi-functional industrial, experimental, reading, learning and teaching aims.

Chapter 1 deals with hydrocolloids as adhesives. The adhesive properties of many hydrocolloids (gums) have been known for centuries. The word 'gum' means a sticky substance, and was previously defined as such by the Egyptian term *qemai* or *kami*, referring to the exudate of the acanthus plant and its adhesive capacity. A large number of hydrocolloids have been mentioned in the literature as adhesive agents as have some lesser-known water-soluble gums which possess wet-adhesive bonding properties. Nevertheless, aside from two chapters and a few manuscripts written by this author, there is almost no detailed information on this important topic, in particular with respect to its connection to the food industry. The chapter provides a definition of adhesive materials, a historical overview, details on natural and synthetic hydrocolloids used for this purpose, creating multi-layered foods by adhesion, hydrocolloid adhesion tests, hydrocolloids as wet glues, physical properties of hydrocolloid glues and their dependence on layer thickness, moisture content and molecular weight.

Chapter 2 deals with hydrocolloid coatings. Most of the available information on food coatings focuses on moieties created by drying hydrocolloid solutions or blends. This chapter discusses coatings created by drying hydrocolloid gels, solutions (this part is similar to what can be found in other books), blends and wax–hydrocolloid mixtures. Never have this many different coatings been reviewed separately in the literature, even in books dedicated solely to coatings. In addition, gloss has become a forgotten food property and the way in which food surface gloss can be changed by coating films is discussed here. The chapter includes discussions of various coatings of the past, present and future, methods of testing coatings, coatings for meats, seafood, cheeses and other foods. Special emphasis is given to the topic of how to design a hydrocolloid coating from both its practical and theoretical aspects.

Chapter 3 deals with liquid-core hydrocolloid capsules and hydrocolloid macro-capsules. Only limited information can be found in the literature pertaining to liquid-core capsules and then most often from a medicinal rather than a food point of view. The chapter contains a definition of liquid-core capsules, discusses membrane formation, mechanical properties of capsules, sizes, formation methods, food and other uses and some information on dry macro-capsules.

Chapter 4 deals with multi-layered hydrocolloid products. Sporadic information on multi-layered foods can be found in the literature. However, no attempt has been made to summarize all the information on multi-layered (multi-textured) products, which are nevertheless important for their ability to convey the sensation of eating many textures at once. The chapter includes information on multi-layered foods in general and in the Orient, explains why such products are desirable, how to glue gel layers, mechanical properties of such foods and theoretical approach of how to predict their properties.

Chapter 5 deals with flavor encapsulation. This topic has been reviewed before, but our emphasis is on the role of hydrocolloids in the preparation and possible uses of new gums within such products. The chapter includes a description of spray-drying and extrusion processes for flavor encapsulation, hydrocolloids as suitable matrix-builders, their performance in flavor encapsulation and flavor retention.

Chapter 6 describes immobilization for food purposes. Most of the information in the literature covers the biotechnological uses of immobilization. Here a trial has been made to emphasize the role of food. The chapter defines immobilization and its techniques, describes structure, texture, physical and chemical properties of the preparations, procedures to modify structure and performance, stability of preparations followed by many examples and future uses in food science and biotechnology.

Chapter 7 describes the texturization of vegetative materials. This chapter contains a description of the role of hydrocolloids in preparing gum-based foods, novel structured food vegetative products, the role of particle size and properties in the mechanical properties of texturized fruits, alternative processes for preparing fabricated fruits, how to improve the succulence of such products, multi-layered texturized fruits and the trend towards the broadening of using such products where a major component of its content is of vegetative source.

Chapter 8 emphasizes the food uses of hydrocolloid cellular matrices. The chapter contains a definition of cellular solids and describes different structures of cellular solids. It gives examples of edible cellular solids; the unique nature of hydrocolloid cellular solids, different ways of producing hydrocolloid cellular solids, such as physical, chemical, slow fermentation, enzymatic procedures, combining preparation procedures, control of porosity and its relationship to slow release, producing cellular solids by immobilization, inclusion of oil in cellular solids, other inclusions and their influence on mechanical properties, food and other uses of hydrocolloid cellular solids.

Chapter 9 deals with the uses of hydrocolloids for producing special textures. This chapter emphasizes special textures and tries to avoid more established ones. The chapter includes a description of processed foods, restructured foods, hydrocolloid combinations, types of fabricated foods, molded products, uses of hydrocolloids in spray-dried products, stabilization, frozen products, freeze texturization, porous products and others.

This book has been written during the last two years. It is dedicated to the blessed memory of my late father, Eliezer, who introduced me to the magic of the written word and its importance and eternity. The book contains a description of many non-traditional uses of hydrocolloids that were developed in many hydrocolloid research and development laboratories all over the world, including ours. I hope that this book will assist all levels of readers, from the student to the food scientist and help many other researchers, engineers and industrialists in their search for different yet unexplored uses and applications of hydrocolloids in food and other areas. Their future comments and questions will be very much appreciated.

Acknowledgements

I wish to thank the publishers for having accepted this book for publication. Special thanks to Mr. Nigel Balmforth for the efficient way in which he contributed to the formation and processing of this manuscript. I wish to thank my editor Camille Vainstein for working shoulder-to-shoulder with me when time was getting short. The help of my colleague and friend Mr. Omri Ben-Zion supporting me with literature research, references and good advice is very much appreciated. The efficient help of Mrs. Hanna Ben-Or with completion of so many old or non-accurate references was unbelievable. The contribution of pictures for the book by Mr. O. Ben-Zion and my previous students and present colleagues Drs. Varda Hershko, Yossi Tal and Nir Kampf and the permissions we got from different publishers are very much welcomed. The love and patience of my family members Varda, Ya'ara, Eran and Yoav was very supportive during the difficult last years when my father was fighting for his life and we were under huge pressure from different directions. The support I got from my teacher and the famous scientist Professor Micha Peleg of the University of Massachusetts during these and previous years cannot be covered by a few adjectives and will be appreciated forever. Last but not the least, I wish to thank the Hebrew University of Jerusalem for being my home and refuge during the last ten years of very extensive research and teaching.

Amos Nussinovitch
Rehovot, Israel, October 2002

About the Author

Amos Nussinovitch was born in Kibbutz Megiddo, Israel. He is the son of holocaust survivors. Nussinovitch served as a combat soldier in the October (Yom Kippur) and Lebanon wars and his heritage and the horrors of war influenced and changed his life, thoughts and attitudes. He studied chemistry at the University of Tel-Aviv, and food engineering and biotechnology at the Technion-Israel Institute of Technology. Nussinovitch worked as an engineer in several companies and was involved in several research and development projects, in both the United States and Israel, especially in studying mechanical properties of liquids, semi-solids, solids and powders. Currently, Professor Nussinovitch is the director of the Biochemistry and Food Science Department at the Faculty of Agricultural, Food and Environmental Quality Sciences of the Hebrew University of Jerusalem. Professor Nussinovitch is the leader of a large group of researchers working on theoretical and practical aspects of hydrocolloids, including: coating of cells and foods, special glues, water-soluble polymer uses in paper, hydrocolloids in cosmetics, explosives, ink, and special cellular solids and biological carriers. He is the author of the book *Hydrocolloid Applications*, has written numerous papers on hydrocolloids, and physical properties of foods and has about 30 patent applications. This book is especially devoted to novel applications of hydrocolloids in foods.

Chapter 1

Hydrocolloid Adhesives

1.1 Introduction

Every discipline develops its own specialized vocabulary, and the field of adhesives is no exception. According to the ASTM definition, an adhesive is a substance capable of holding materials together by surface attachment. In a bonded structure, the bond (or joint) is the location at which two materials, termed adherents, are held together with a layer of adhesive. The forces of attraction acting across the adhesive/adherent interface are, therefore, those responsible for holding the materials together (Hartshorn, 1986). They may arise via the chemical-bond formation, physical interactions such as dispersion forces, or mechanical interlocking. In the last, the adhesive is assumed to penetrate pores and any other surface irregularities of the adherent, which are always present on a microscopic level, thus producing good mechanical interlocking (Hartshorn, 1986).

The adhesive properties of many hydrocolloids (gums) have been known for centuries. The word *gum* means a sticky substance, and was previously defined as such by the Egyptian term *qemai* or *kami*, referring to the exudate of the acanthus plant and its adhesive capacity (Glicksman, 1982). A large number of hydrocolloids have been mentioned in the literature as adhesive agents (Bauman & Conner, 1994; Chen & Cyr, 1970). They include gum talha (similar to gum arabic), gum ghatti, gum karaya, gum tragacanth, arabinogalactan (AG), dextran, pectin, tapioca-dextrin, carboxymethyl cellulose (CMC), methyl cellulose (MC), hydroxypropyl cellulose (HPC), hydroxypropylmethyl cellulose (HPMC), carbopol, polyethylene oxide, polyvinyl pyrrolidone (PVP), polyvinyl alcohol (PVA), retene, pullulan and chitosan. Some lesser-known water-soluble gums which possess wet-adhesive bonding properties include gum angaco, brea gum and psylium seed gum (Mantell, 1947), gum cashew (Howes, 1949), gum damson, jeol, myrrh (Hagqist *et al.*, 1990) and scleroglucan (Glicksman, 1982).

1.2 Mechanical testing of adhesive joints

1.2.1 Tensile test

The application of tensile loads to adhesive systems should be considered before testing, since the tensile strengths of most substrates are higher than those of the adhesives joining them. Thus failure occurs at the adhesive joint responsible for joining the bonded structure. Tension is used to examine adhesive joints, bonds or design via the application of perpendicular forces on the adhesive layer. Calculations of tensile strength are based on average stress value and not on the non-uniform stress values within the adhesive (Portelli, 1986). Different modes of failure have been recognized: failure within the adhesive layer – cohesive failure; failure at the interface between the adherent and the adhesive – adhesive failure; and failure of the adherent itself. After testing, the ratio between adhesive and cohesive failure should be estimated. Samples in which the failure occurs within the substrates should be discarded, since they do not constitute a test of the adhesive material (Portelli, 1986).

1.2.2 Shear test

The structural adhesive within bonded structures is usually designed to sustain shear loads most of the time, the reason being the higher strength of adhesives in shear, as compared to that under peel or tensile loads. Applied loads act in the plane of the adhesive layer when the adhesively bonded structure is considered under shear. As a result of the loads, the adherents can slide, which in turn causes shearing or sliding of the adhesive. A certain length of overlap between adherents is typical when lap-shear

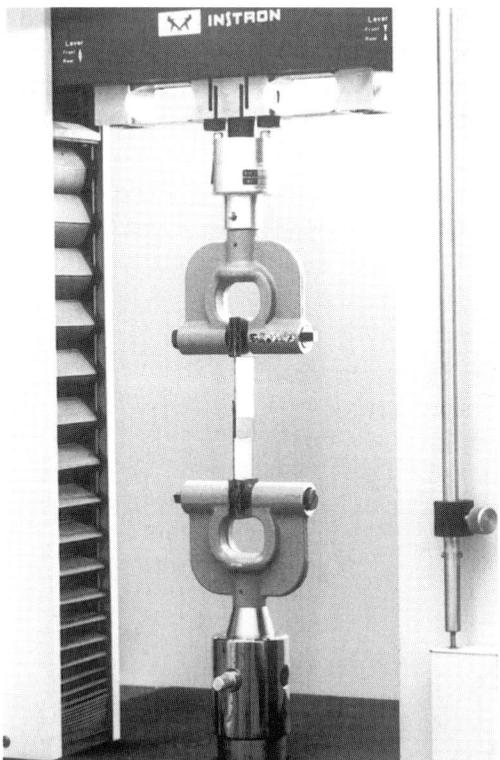

Fig. 1.1 Specimen mounted on an Instron universal testing machine during the lap-shear test. (From Ben-Zion & Nussinovitch, 1997a, with permission from Elsevier Science.)

tests are to be applied (Fig. 1.1). The layer between the overlapping area consists of the structural adhesive, and tension-to-failure is used during the lap-shear test. The critical shear strength is calculated by dividing the load at failure by the area of the overlap (Portelli, 1986). Tensile forces can be replaced by compressive ones, when compressive shear tests are applied to the sample until failure. Such loading could be beneficial when better alignment of the adhesive bonds is achieved in parallel to a reduction in the non-uniformity of the stress (Portelli, 1986). The shear strength and modulus of the adhesive can be estimated by torsional shear stress. Sample alignment is very important for achieving more homogeneous stress distribution. Other important parameters include yield stress, thickness and other characteristics of the adherent (Portelli, 1986).

1.2.3 Peel test

Peel tests are used for quality control or for comparison of different adhesives, and are important in quantifying stripping or peeling forces. In peel tests, flexible adherent that is adhesively bonded to a rigid or flexible adherent is stripped. During this operation, the distribution of the stress within the peel joints is complex and is influenced by the properties of the adherent and the geometry of the joint. The width of the peeled adherent is a major factor in calculating the peel strengths, usually in pounds per inch. The average load needed to maintain the peeling after initiation is used to estimate the peeling force (Portelli, 1986).

In most cases, adhesion of a material is estimated by peel testing (Fig. 1.2). In such simple tests, the rate of failure is easily controlled. The method can suffer from bending during the peeling test and from the effects of a material's plastic behavior. The quantification of peeling force is dependent on the energy involved in treating the coating. In peel tests, a part of the energy dissipates during the testing (Croll, 1983). When a coating is peeled at 90° to the substrate, an energy-balance concept is used; i.e.

$$\frac{F}{b} = \gamma - t_c U_R \qquad (1)$$

where F and b are the peel force and width of the coating, respectively, and F/b is a measure of the adhesion strength, i.e. the force per unit width necessary to peel the

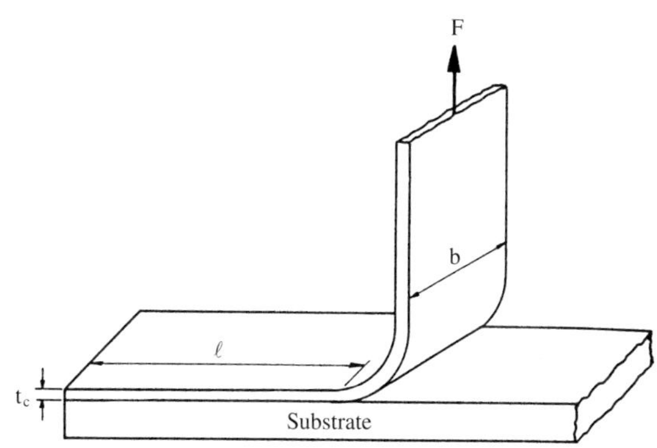

Fig. 1.2 Peel test configuration. (From Croll, 1983.)

coating. Other parameters in this equation are the thickness of the coating – t_c, the recoverable strain energy – U_R and the interfacial work of adhesion – γ (Croll, 1983).

1.2.4 Other tests

In general, the aforementioned three tests can be regarded as static tests. In other words, they are conducted in a matter of minutes (short periods) at constant loading rates. Other tests include dynamic loading, creep, impact and fatigue, and they can be of long (i.e. hours, weeks or months) or short (i.e. seconds) duration. Long-duration tests are important in determining the damage tolerance of adhesives and their long-term durability (Portelli, 1986).

1.3 Hydrocolloids as adhesive materials in foods

Gums can be used in bioadhesive applications in medicine and cosmetics (Bottenberg *et al.*, 1991; Bouckaert & Remon, 1993; Irons & Robinson, 1994; Robinson *et al.*, 1987; Smart, 1991; Smart *et al.*, 1984): for example, adhesive bioelectrodes (Keusch & Essmyer, 1987), cosmetic preparations (Toulmin, 1956), pressure-sensitive adhesives (Kiyosi & Yasuo, 1986; Piglowski & Kozlowski, 1985), ostomy rings (Glicksman, 1982), adhesive ointments (Kanig & Manago-Ulgado, 1965) and dental adhesives (Shay, 1991). They are also used in the paper, wood and leather industries (Bauman & Conner, 1994; Sadle *et al.*, 1979; Torrey, 1977). Tobacco is sometimes packaged in paper cylinders that are glued together with starch or gum tragacanth (Brief, 1990). Gums (apple pectin and gum arabic) and sometimes mixtures of gums and aqueous sugar solutions are used as adhesives and edible glues in the food industry; they can also contribute to viscosity and thickening, as well as to potential gelation acceleration (Mazurkiewicz *et al.*, 1993).

1.3.1 Crumb and batter adhesion

Better eating quality and an enhanced variety of options within foods can be achieved by the common practice of coating with breadcrumbs, batter or other compounds. Suspending flour in water produces batters, which can also include various thickeners, salt, sugar, coloring and flavoring materials. Batters are applied to various products' surfaces, including potato, fish and poultry. Tempura is a viscous batter producing a single layer around coated products that are not breaded. A low-viscosity batter suspension with adhesive qualities can be used to enrobe food products before breading is performed (Fellows, 1990). During manufacturing, the food can be forwarded from a battering machine to a breadcrumb applicator. These machines use different specific-particle-size breadcrumbs that are fragile, flavored and colored for a predetermined purpose. The product is submerged in the crumbs or passed through curtains of batter. Gentle pressure is used to achieve better attachment of the breadcrumbs to the product, and air blowers are used to remove the excess (Fellows, 1990).

Many kinds of crumb crusts can be made from graham cracker, ginger snap, vanilla wafer and melba toast crumbs. Crumb crusts always require a pre-cooked filling, and these fillings should be fairly dry. Crumb crusts are ideal on the bottom of square and rectangular pans as a base for pie-type fillings. A portion of the crumb mixture can be reserved to sprinkle over the top surface of the filling for decoration (Gates, 1981). Foodstuffs such as fish or meat can be coated with a wheat flour composition containing an emulsifying fat, crumbs and a natural gum or adhesive agent prior to frozen storage (Nippon Suisan Kaisha Ltd., 1973).

Gums are commonly used to adjust variations in cold-batter viscosities that occur due to other ingredients (e.g. starches, flour) in the batter. Starches are added to batters at levels of 5–15% (w/w, flour basis). Most food gums are used in the 1–3% (w/w) range (i.e. 0.5–1.0% w/w, final wet batter basis); thus their use can be more economical (Meyers, 1990). Gums are used to suspend batter solids, particularly in low-solid systems. Stabilization is achieved via the thickening and interfacial activity of the gum. The latter enables stability of spice oleoresins in water. The yield value (initial resistance to flow under stress) of, for example, xanthan and gum tragacanth, enables a suspension of heavy particles at low gum concentrations (0.10–0.25% w/w). Gums' ability to form gel or film moieties (see Chapter 2) provides strength and flexibility to the batter system, which in turn provides adhesion to prevent *blow-off* and cohesion to prevent *pillowing* (Meyers, 1990). Thus the resistance to defects caused by handling of breaded and battered products increases. Gums such as MC and HPMC improve adhesion strength not only because of their film-forming ability but also because of their thermal gelation properties. In general, the adhesion strength supplied by gums increases with gel strength. Adhesion also increases with gum concentration up to a point; at this point it begins decreasing due to steam pressure buildup. The adhesive strength also increases with increases in the molecular weight and viscosity of the batter. Batters with good adhesion and cohesion can be achieved by using pre-dust (Kuntz, 1997) or pre-dip systems composed of 80% adhesion starches (modified starch treated with ethylene oxide or sodium hypochlorite) (Campbell, 1972); or proteins and protein sources (e.g. whey, egg albumen, gelatin, wheat gluten, non-fat dry milk, soy protein concentrate, sodium caseinate) and 20% of a suitable gum. However, the adhesion properties of the batter and gum system are also influenced by other properties of the substrate such as shape, surface area, ice-crystal layer or ice glaze and moisture (Meyers, 1990).

Dry blending is the preferred method of incorporating gum into a batter system. Batter/breading companies blend dry ingredients with gums, and the batter is prepared by adding water to an appropriate mixer. Gums can also be suspended in the water phase, in oil or in high-fructose corn syrup for subsequent hydration with water. Three other application modes involve direct pre-dusting of the substrate surface with gum and other batter ingredients; pre-dip of the surface in a batter coating solution; and using the gum as a coating or incorporating it into the breading portion for better adhesion and barrier properties (Meyers, 1990). Pre-dusting has also been reported to create uniform adhesion and a smooth coating, and the use of high-amylose starch in the batter contributes to crispier products (McGlinchey, 1994).

A few considerations should be taken into account when choosing a gum to serve in a batter system. In a high-solids system, gum hydration before the addition of these solids is essential. In the case of ionically charged gums, which can be less effective due to the presence of salts, additional gum may be needed to reach the required effectiveness (Meyers, 1990). Using batter viscosity as a quality control indicator for batter pickup is sometimes misleading; the actual pickup should therefore be monitored. The special features and functions of gums make them very attractive in batters. These properties include the thermal gelation and film-forming properties of the cellulose derivatives MC and HPMC; the ability of xanthan, CMC and guar gum to adjust batter viscosity at low concentrations; the compatibility of carrageenan with proteins and its ability to stabilize and improve the gel strength; the ability of gum ghatti to thicken batters at low cost; and the influence of alginate to promote starch gelatinization, which in turn promotes adhesion and barrier properties (Meyers, 1990).

Several reports have dealt with different aspects of crumb and/or batter adhesion by gums. These reported adhesive materials were dextrin and soluble alginate salts. A direct relationship between breadcrumb stickiness and the qualitative and quantitative distribution of dextrins in bread made from sprout-damaged wheat of flour containing added bacterial, fungal or cereal α-amylase was reported (Every & Ross, 1996). A process for coating foods with oil-saturated crumbs has been described. A batter containing a soluble alginate salt is applied to the food product as an adhesive for the subsequently applied crumbs which are then also applied. Subsequent spray- or curtain-coating with a polyvalent metal ion solution results in the formation of an insoluble alginate gel which holds the crumbs in place so that oil may be applied by spray or dip without crumb loss or contamination of the oil (Anon, 1979). Another breadcrumb composition which adheres to a moistened comestible, e.g. chicken pieces, pork chops, fish fillets or vegetable strips, during coating and cooking without the necessity of batter coating has been described. The formulation includes breadcrumbs, whose particle size is such that most, if not all, of the crumbs are retained on a 20-mesh US standard screen after passing through a 5-mesh US standard screen, and an adhesive, applied to the surface of the crumbs. The adhesive contains a protein that is marginally more than 1% by weight of the crumbs, and (optionally) a starch and/or a gum. An example of such a composition contained (g): bread crumbs (75), egg white solids (10), gum arabic (0.9), water (17.1) and a spice blend (10) (Rispoli *et al.*, 1987).

Another report deals with the importance of starch in battered food production. It is discussed with reference to its use as an ingredient in batters and breadings. Aspects considered include: pre-dusting with starch to create uniform adhesion and a smooth coating; control of batter viscosity for uniform pickup of solids; use of high-amylose starch in the batter to create crispier products; and use of lower amylose starches for battered vegetable products (McGlinchey, 1994). The use of batters (tempura and fish and chip batters) on seafoods and methods to improve their adhesion, such as the use of starch-based pre-dusts and application of gums, are discussed elsewhere (Kuntz, 1997). A batter-mix starch having consistently high adhesion characteristics is prepared by oxidative treatment of commercial starch with a protein content of greater than 0.7% by weight (Campbell, 1972).

The application of batters and breadings to food substrates is a continual challenge to food producers (Suderman, 1992). A limited number of methods are available to estimate the adhesion degree of batter or breading. An earlier method (May *et al.*, 1969) suggested weighing battered products after placing them in a container with water and agitating with compressed air for about 15 min to remove the batter. The product is then blotted for 2 min to remove excess moisture and reweighed. The percentage of breading is then calculated (May *et al.*, 1969). Ten years later, the adhesion of coatings to food products was evaluated by a quicker method, lasting ~1 min (Suderman & Cunningham, 1979). The process consists of placing a breaded product on a standard wire sieve in a portable shaker, which is turned on for 1 min. Following this operation, the breading crumbs that accumulate in the catch pan are weighed on an analytical balance. Percent loss is calculated from the weight of the breading crumb, divided by the outcome of product weight with pre-dip and breading minus the towel-dried product's weight (Suderman & Cunningham, 1979).

1.3.2 Adhesion of breading to shrimp and fish

Shellfish include crustaceans and mollusks. Shrimp may range from greenish-gray to brownish-red when raw, but when cooked they are always reddish-white. Shrimp are usually sold with the heads removed. Two pounds of shrimp in the shell yield one pound of peeled, cleaned shrimp. Thus, it is more economical to buy peeled shrimp if the price per pound is less than double that of shrimp in the shell, and labor is saved as well (Gates, 1981).

In the past, coating of seafood with batters and breadings was probably very simple (Suderman, 1992). Batters were composed of flour, seasonings (salt and other spices) and water for hydration. The early breadings made use of dried breadcrumbs. An egg wash, or milk or water pre-dip was used to adhere the batter (Suderman, 1992). Today breadings undergo baking and further processing to achieve the required shapes, sizes, flavors and textures. *Japanese-style* breadings and *homemade* commercial breading are produced on a large scale (Quigg, 1980). The quality of battered and breaded seafood products is largely dependent on the quality of the starting substrate, the coating system, the use of starch to adhere the coating, the machinery and other factors that are discussed briefly later on in this chapter (Suderman, 1992; LaBell, 1994). Adhesion problems during breading of shrimps were overcome by using a vegetable gum pre-mix instead of guar gums. Breading (dry toasted crumbs or Japanese bread crumbs) is applied both manually and mechanically. Shrimps are peeled, deveined, washed, inspected and pre-dusted prior to battering with gum and breading. The vegetable gum blend improves adhesion during breading and holds the breading on the shrimp during freezing and deep-fat frying (Toloday & Andres, 1975).

Fish blocks, the starting material for battered or breaded fish portions, are cut into the desired width and shape. An ice glaze (a thin layer of water left on the cut surface and frozen by the fish's temperature) results in poor batter adhesion to the fish portions. In fact, batter *blows-off* the fish surface. Adding salt or increasing the salt content of the product helps melt the ice glaze (Suderman, 1992). If the mixing time is not sufficient, it results in a partially hydrated batter that contains lumps of dry matter and suffers from inferior adhesion. A thick batter (though not excessively so) results in good adhesion. If batter does not adhere properly to the seafood, part of the coating blows-off during frying. Thin batters suffered from not having internal cohesiveness, resulting in poor bonding with the fish surface (Suderman, 1992). Pre-dusting of seafood surfaces improves batter adhesion. Following pre-dusting, a too rapid application of wet batter reduces the quality and quantity of batter adhesion. The frying oil can influence the batter or breading adhesion, by partially stripping excess batter. Therefore, extra attention should be paid to eliminating voids (bare areas that have no adhesive bonding), as these allow the oil to penetrate under the batter and blow it away (Suderman, 1992). In conclusion, good seafood-coating adhesion depends on both the substrate and the coating's properties. However, generally speaking, not much scientific literature on seafood-coating adhesion exists (Suderman, 1992).

1.3.3 Adhesives for snack foods

Snacks provide an immense competitive market for its producers. Some of these products are consumed by children and have new shapes and flavors that are even exaggerated by the producers to lure more consumers. Therefore, a few of these products look strange to the adult eye (Sutherland *et al.*, 1986). Potato-based snacks (e.g. potato crisps, cheese-flavored potato snacks) may be made from potato pieces or from reformed potato powder. As raw material, potatoes of the correct sugar content must be used. Cooking regularly consists of frying. The commercial backgrounds and technologies involved in producing cereal-based snacks and potato-based snacks have some similarities (Sutherland *et al.*, 1986).

The issue of low-fat snacks is an important one. Therefore, a hot gum solution that can be brushed onto the dough surface and contribute to its taste without making it sticky and without changing its caloric value is a valuable product to aim for. The problem is compounded by the fact that removing fat from sweet dough results in a bread-like texture. Tic Gums® of Maryland has launched an adhesive for snack foods that adds no fat to the product. This cold-water-soluble non-fat snack adhesive is composed of gum arabic and several other hydrocolloids and solids, including maltodextrin, and is available in powder or liquid form for the adhesion of seasonings to snacks (e.g. pretzels, tortillas and extruded snacks). The product is sprayed with adhesive, tumbled with spices and dried for 5 min. The process adds no appreciable caloric value and no fat to the snack and therefore negates the need for antioxidants (Anon, 1993). Removing fat from a sweet dough product is problematic in that a bread-like texture results when no laminating fats are used. However, lamination with a gel-forming hydrocolloid will create an open texture with layers that resemble those of a fat-laminated pastry. Gellan gum is easily used in this application because it can be held as a hot solution and gels quickly after being applied to the dough. In trials, a 3% solution of gellan gum was made by dispersing 3% gellan gum and 0.2% sodium citrate (sequestrant) in tap water, and heating the mixture to 90°C. The hot gellan solution is brushed on the dough during the lamination step as if it were melted fat. The gel sets quickly and the dough is folded, rolled and laminated further.

The water from the gellan gum solution adds moisture to the dough without creating excess stickiness. The resulting pastries exhibit lamination, and are moist and flavorful. The added water also helps maintain product conformity during processing. Over-mixed dough laminated with a hydrocolloid solution consistently produced higher-volume pastries. A leavening agent could be added to the hydrocolloid solution to open up the structure further. Other gelling agents should work as well if they set rapidly. Egg white may have a similar function. However, non-gelling solutions may add too much water to the dough, making it sticky (Valli *et al.*, 1997).

1.3.4 Poultry patties and nuggets

Since the end of World War II, most poultry, particularly chickens and turkeys, in industrialized countries has been produced using large-scale agricultural and processing techniques. As a consequence, the relative cost of poultry has fallen markedly in relation to other meats and what was once considered to be a luxury item is now a cheap staple food (Sutherland *et al.*, 1986). Successful application and adhesion of batters and breadings to poultry depends on many factors:

- Scald temperatures may be significant in affecting adhesion to poultry skin, since a high scald temperature removes the skin's cuticle.
- Removing the cuticle results in better batter and breading adhesion (Suderman, 1992).
- Bird age does not noticeably affect the skin ultrastructure.
- Freezing poultry parts may slightly improve the adhesion.
- Breading adhesion and uniformity were found to be progressively lower with precoating dips of milk, water and no dip, in that order.
- Increased batter viscosity contributes to better breading pickup.

Adhesion of batter and breading increased when CMC was used instead of guar, gum tragacanth or xanthan (Suderman, 1992). However, different conclusions as to gum's influence on batter adhesion were reached in the research conducted on chicken nuggets (which did not report exact raw materials, formulations or procedures as in the former reference). This study assumed that improved adhesion of batters to food products such as poultry patties and nuggets would further reduce their cost and improve the product quality. Time-dependency, apparent viscosity, shear thinning behavior, recovery and adhesion characteristics were determined on 30% solid flour-based batters containing hydrocolloids (guar, xanthan and CMC at 0.25, 0.5 and 1.0%). Most batters were thixotropic. Batters containing xanthan gum had the greatest apparent viscosity followed by, in decreasing order, guar gum, CMC and control batters. Apparent viscosity showed a high positive correlation with batter adhesion characteristics measured on chicken nuggets as coating pickup, overall yield and cooked yield. Mixer viscometry techniques were useful for tracing changes in rheological properties caused by mixing speed and time (Hsia *et al.*, 1992).

Changing gums and protein sources within the mixes can control adhesion of breading mixes to poultry skin. Such proteins can be gelatin, egg albumen, soy, whey and non-fat dry milk. The different water-soluble polymers are sodium CMC, guar, tragacanth and xanthan. Egg albumen and gelatin were found to be superior to other tested proteins in terms of adhesion enhancement. CMC was favored over the other gums. Increased levels of hydrocolloids and proteins did not contribute to the adhesion, perhaps because an asymptotic contribution was reached at lower concentrations (Suderman *et al.*, 1981). As in other cases, the adhesion of batter and breading is dependent, among many other things, on the surface properties of poultry skin. Growth of microorganisms on poultry skin did not affect breadcrumb loss. The breading mix contributed to maintaining product tenderness and reducing the weight loss incurred during baking. Changing the pre-dip from water to milk or evaporated milk also reduced volume losses due to baking (Suderman, 1980). Another report described a dry coating mix composed of water, flour, starch, gum (0.1–1.0%) and oil formed into particles. These particles were fried in oil at 160–227°C, excess oil was removed, and the other coating ingredients were mixed in. These mixes can be applied to various meat and vegetable pieces prior to baking, to achieve better taste, texture and appearance (Rispoli *et al.*, 1987).

1.4 Multi-layered gels and texturized fruits

1.4.1 Multi-layered gels structured by adhesion

A simple way of achieving different textures and tastes in the same bite is to construct a food product made up of different layers. A few such multi-layered food products are already on the market, for example crunchy wafers that include a sweet vegetable-fat-based chocolate or vanilla taste filling between brittle wafers. For children, a multi-layered, sweetened, agar-based confection can also be found. The texture of its layers is similar, but their tastes and colors can differ. In the Orient, where awareness of different gel textures is much more developed than in the West, a curdlan-based, sweetened, multi-layered gel has been developed (Harada, 1977, 1979; Ikeda *et al.*, 1976). In this case all layers are built from the same hydrocolloid, because two types of curdlan gel can easily be prepared from its powder by heating the suspension to different temperatures. Multi-layered foods based on hydrocolloids are important in the framework of foods of the future (Nussinovitch, 1997). One publication describes a gelatin dessert product which produces a clearly demarcated, two-layered, two-colored dessert when dissolved in hot water and then chilled. Typical flavors include chocolate, coffee and a variety of fruit flavors (Kadison & Scheiner, 1985).

The stiffness of the layered array or product can be estimated from the deformability moduli of its individual layers. An empirical mathematical model to predict the deformability modulus of a multi-layered gel array was successfully applied to a series of double-layered gels composed of agar and one of four galactomannans [locust bean gum (LBG), guar gum, tara gum or fenugreek gum] in three different layer-thickness combinations (Ben-Zion & Nussinovitch, 1996). A multi-layered gel was constructed using a previously described adhesion technique where hot solutions of a suitable hydrocolloid or hydrocolloid mixture are poured on already gelled layers of identical or different compositions at room temperature. This operation yields a multi-layered gel structure (Fig. 1.3). The model is based on the assumption that the uniaxial stress in the layers is the same and that their deformations are additive. A typical example of the stress–strain relationship of gels having the same composition and three different layer-thickness ratios is presented in Fig. 1.4. It should be noted that as the proportion of the agar layer (thickness) increases within the gel, the deformability modulus of the whole array increases. This is demonstrated in Fig. 1.4,

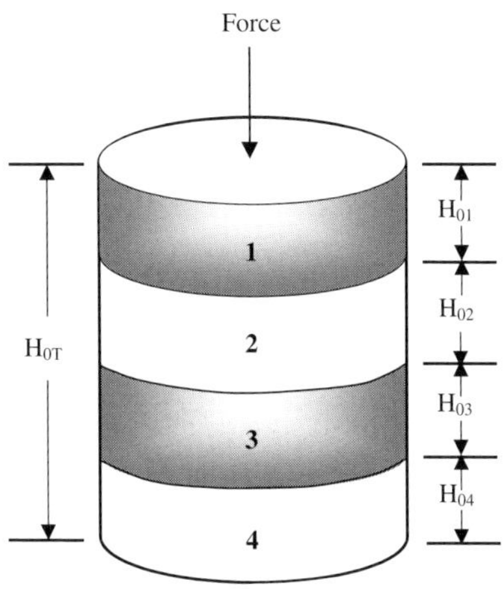

Fig. 1.3 Schematic view of the geometry of a multi-layered array of gels. The H_{0T} values represent the initial thickness of the whole array.

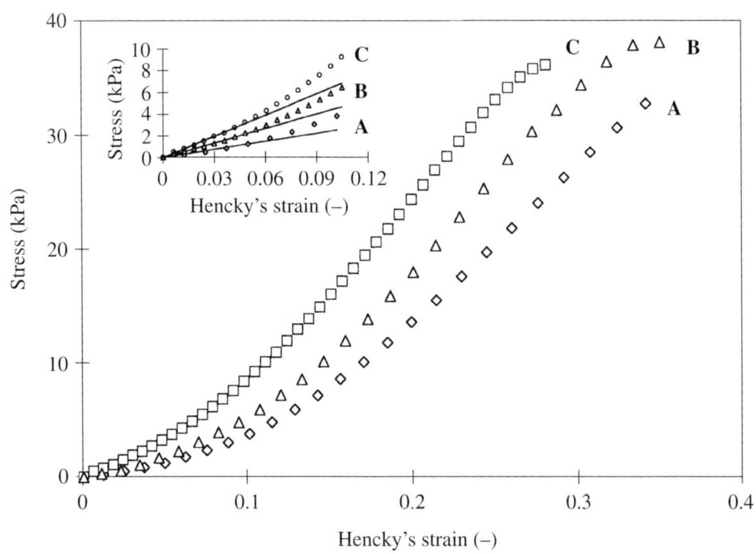

Fig. 1.4 Stress–strain relationship of double-layered gels made up of agar and agar/galactomannan. The thicknesses of the separate layers are: A = 33:67; B = 50:50; C = 67:33. The inset illustrates a comparison between the experimental and predicted stress–strain relationships. The solid lines represent the fits of the mathematical model. (From Ben-Zion & Nussinovitch, 1996.)

where the slope of the curve rises most steeply when two-thirds of the specimen is composed of agar. The type of galactomannan combined with the agar had a significant effect on the mechanical properties of the resultant multi-layered gel. Inclusion of galactomannans increased the adhesiveness between the layers, resulting in a finely glued multi-layered product. The higher the degree of substitution within the galactomannans, the stiffer the multi-layered gel formed. No significant differences were found between deformability moduli calculated from experimental results and those predicted by the model. The model provides a tool for estimating multi-layered gel stiffness and may be applicable to other food systems that behave similarly (Ben-Zion & Nussinovitch, 1996).

1.4.2 Different adhesion techniques for multi-layered gels

A few techniques can be used to *glue* the layers of a multi-layered gel together. The first simply takes advantage of the following phenomenon: pouring hot pre-setting solutions of hydrocolloid or hydrocolloid mixtures on already-gelled layers with identical or different compositions at room temperature will usually (though not always) produce a multi-layered gel. Another gluing method (Fig. 1.5) makes use of two pre-gelled layers of identical or different gels. After smearing the surface of one layer thoroughly (using a fine brush) with agar solution at a temperature of 95°C, the two layers are pressed together. After a short time, the layers adhere strongly. The third method describes the simultaneous preparation of two solutions, followed by pouring them together into an appropriate receptacle. After a short time the layers separate, while the solution is still hot. After gelation, a two-layered gel system is observed. In one example of this method, two different gels could be produced. The first consisted of a bottom layer, made up of konjak mannan and κ-carrageenan. The bottom layer of the second gel was made up of konjak mannan with calcium carbonate and sodium acid pyrophosphate. Full details on dissolution during preparation can be found in Ben-Zion and Nussinovitch (1997b). Three failure patterns were found for the multi-layered gels. When gluing was performed by pouring a hot hydrocolloid solution onto an already-gelled layer, failure in the weakest layer occurred in agar–LBG and agar–fenugreek gels. Layers separated in the double-layered agar–tara and agar–guar gel combinations. With the simultaneous gluing method, no separation was observed: failure occurred at ~45° to the surface of the gel. With the method that consisted of gluing two gelled layers together by smearing a hot solution of agar between them, separation occurred with the agar–LBG, agar–tara and agar–guar combinations, whereas for agar–fenugreek, fracture occurred in one of the layers. Thus the best adhesion was achieved with the simultaneous gluing method, whereas the smearing method yielded the weakest adhesion. The first method gave intermediate results. Figure 1.6 presents typical experimental deformation–force relationships for the three different gluing techniques and compositions, and demonstrates the excellent fit of the proposed mathematical model:

$$\Delta H_T = \sum \left[\frac{c_i F}{(a_i + F)} \right] \tag{2}$$

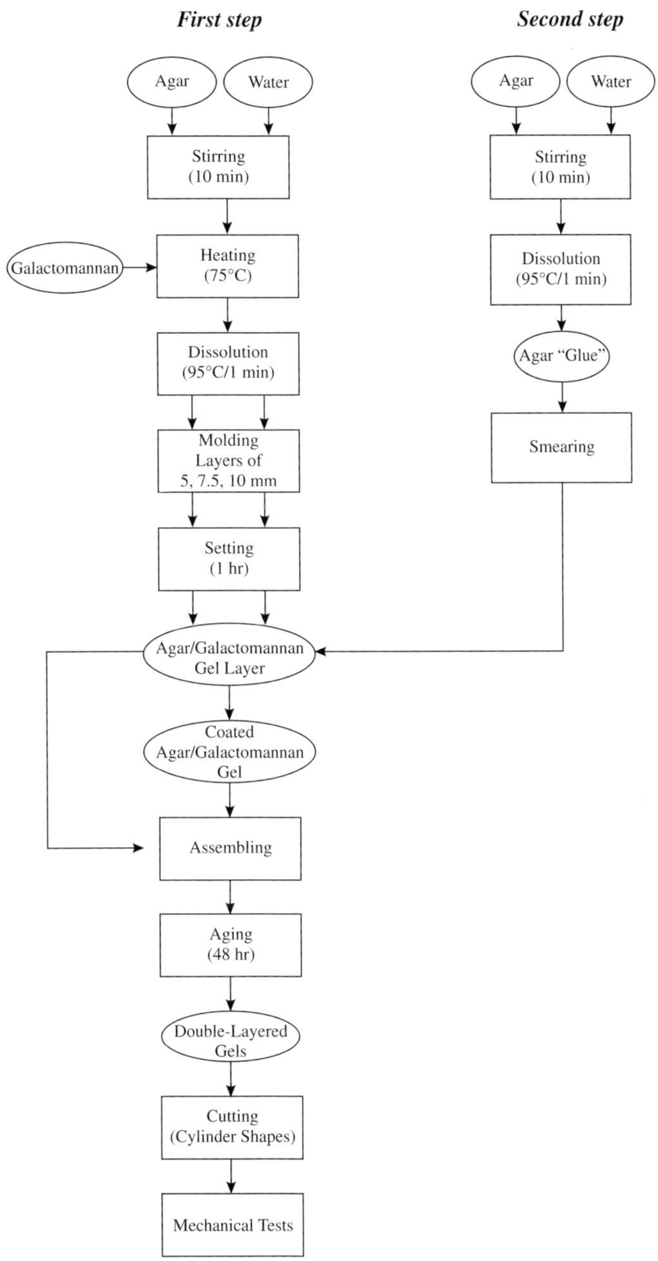

Fig. 1.5 Flow chart of one of the adhesion techniques used: gluing two separate gelled (similar or different) surfaces by smearing agar solution at 95°C onto the surface of one of them, and sticking them together. (From Ben-Zion & Nussinovitch, 1997b, with permission from Elsevier Science.)

where the engineering stress and strain are defined as $\sigma = F/A_o$ and $\varepsilon = \Delta H/H_o$, H_o and A_o being the layer's original thickness and area, respectively; $\Delta H_T = \sum \Delta H_i$; a_i and c_i are constants. It should be noted that the magnitude of these constants may depend on the deformation rate. If the dependency is strong, they should be

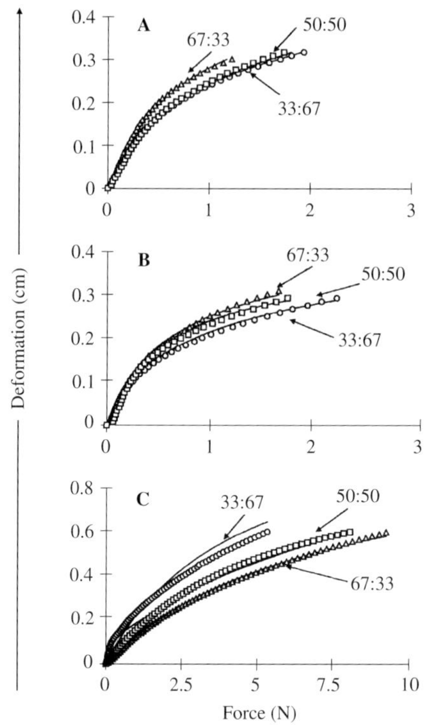

Fig. 1.6 Compressive deformation versus force relationships of double-layered cylindrical specimens (15 × 15 mm). Thickness ratios are noted on the figure. Symbols denote experimental data and the solid lines represent the model's prediction. A–C are related to the gluing techniques mentioned in the text (pouring hot hydrocolloid onto an already-gelled layer, gluing together two gelled layers with hot agar, and pouring two gel solutions simultaneously, respectively). (From Ben-Zion & Nussinovitch, 1997b, with permission from Elsevier Science.)

determined in dimensions and rates that correspond to their deformation as layers. The fitted lines in Fig. 1.6 were predicted using the model with constants calculated from the force–deformation data of the two separate layers. Figure 1.6 contains only three examples of the many combinations used in this work. As stated, very good fit was achieved in all cases, including those where separation between layers was observed at high deformations. The demonstrated predictive capability of this model justifies the use of the assumption on which it was based, i.e. that the rate effects and lateral stress can be ignored in the different multi-layered gel systems tested here, at the reported deformation levels. In principle, the model can be extended to other layered gels with different compositions, using different gluing techniques to adhere the layers. Nevertheless, since the model's success depends on the validity of the assumptions, its applicability always needs to be tested with experimental data (Ben-Zion & Nussinovitch, 1997b).

1.4.3 Adhesion-strength characteristics of double-layered gels

Typical curves for 90° gel–gel peeling are shown in Fig. 1.7. The dashed line represents the calculated peel-bond strength. It is not surprising that whereas experimental

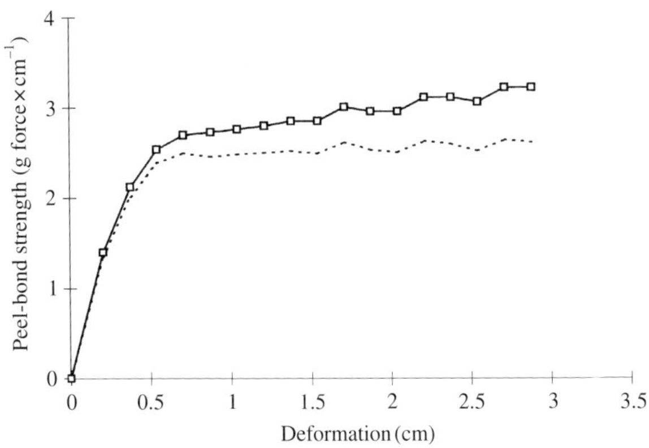

Fig. 1.7 Typical curve for a double-layered gel (2.5% agar with 0.4% LBG) in a 90°-peel test. The dashed line represents the corrected peel-bond strength. (From Ben-Zion & Nussinovitch, 1997c, with permission from Elsevier Science.)

results showed a small increase in peel-bond strength with time (or deformation), the theoretical (calculated) values are fairly constant after almost 3 s, reaching a fairly quick steady state. The combined influence of agar and LBG concentrations on peel-bond strength is presented in Fig. 1.8. It appears that an increase in agar concentration, and in LBG concentration up to 0.6%, increases the peel-bond strength. The presence of 0.6 and 0.8% LBG (which still contributed to the strength of the agar and LBG gel) appeared to cause a decrease in peel-bond strength. In the absence of LBG, the higher the agar gum concentration, the higher the peel-bond strength. The mechanism of adhesion between two identical layers could be explained as follows: when a hot agar or agar–galactomannan mixed gel is being poured onto another similar or identical already-gelled layer and temperatures are higher than those which cause melting, then micro-melting of the *bottom* mono-layer at the interface can occur. Evidence of this phenomenon was observed by Ben-Zion and Nussinovitch (1997b).

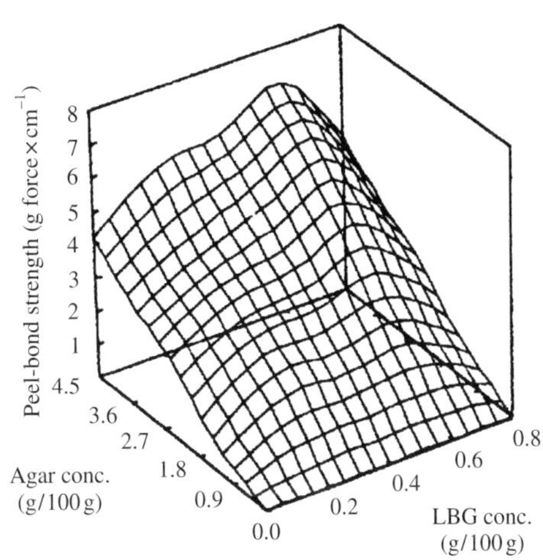

Fig. 1.8 The effect of agar–LBG concentration on 90°-peel-bond strength of double-layered gels. (From Ben-Zion & Nussinovitch, 1997c, with permission from Elsevier Science.)

Galactomannans differ from one another in their mannose : galactose ratio and in the distribution pattern of the galactose residues along the mannan chain. When reheating agar–galactomannan mixed gels above the melting point of the agar, the associations melt, eventually leaving unbound agarose helices (Dea, 1979). Those helices can interact with the galactomannan *smooth regions* which will eventually be present in both layers. In other words, inter-molecular interactions are suggested to be responsible for gel-layer binding. Figure 1.9 shows the influence of different bottom-gel roughnesses on peel-bond strength between identical gel layers. The concentration of the agar inside the layer was varied (1.5, 2.5, 3.5 and 4.5%), and was combined with a constant concentration (0.4%) of different galactomannans (LBG, tara gum or guar gum). A positive linear relationship between peel-bond strength and agar concentration (Fig. 1.9) was found, regardless of the galactomannans used. At all three degrees of roughness, the peel-bond strength between agar–LBG double-layered gels was found to be higher than that between those including tara gum or guar gum. Possible inter-molecular interactions could partially explain the changes in adhesive strength between double gel layers. Peeling strength appeared

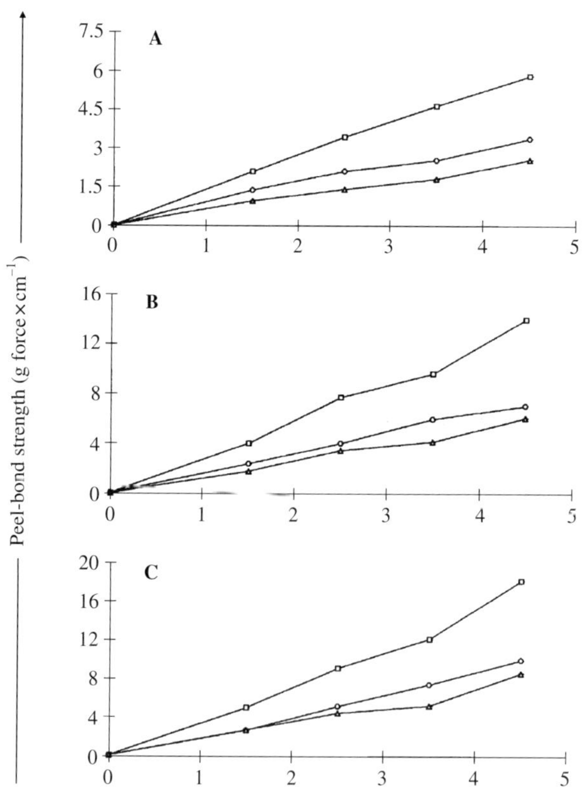

Fig. 1.9 The influence of agar concentration, and galactomannan type (□ = LBG, ○ = tara, △ = guar; all at a constant concentration of 0.4%) and the degree of roughness (A, smooth surface; B, *w* (width) = 0.5 mm, *d* (depth) = 0.32 mm; C, *w* = 0.16 mm, *d* = 1.112 mm) on the 90°-peel-bond strength of double-layered gels. (From Ben-Zion & Nussinovitch, 1997c, with permission from Elsevier Science.)

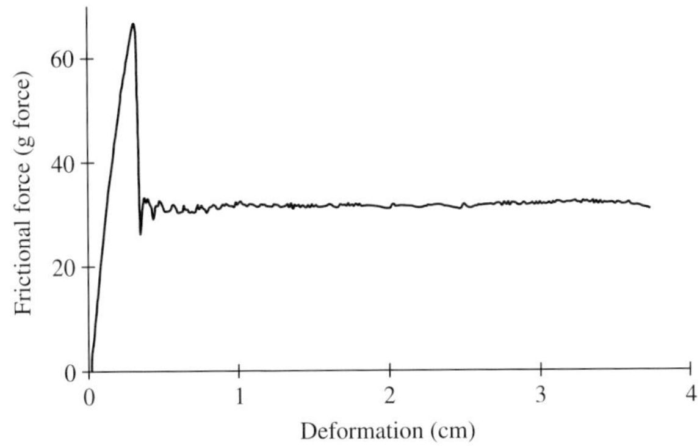

Fig. 1.10 Typical curve of measured frictional force for identical bottom and top gels. (From Ben-Zion & Nussinovitch, 1997c, with permission from Elsevier Science.)

to be higher when the least-substituted galactomannan was combined. The degree of roughness of the bottom-layer surface significantly affected the strength of the bond between any two examined layers. Peel-bond strengths were directly dependent on the degree of roughness, decreasing with decreasing roughness (Fig. 1.9). A friction test can be used to characterize the screened and smooth surfaces of molded gels. In a frictional force versus deformation curve (Fig. 1.10), two typical regions are observed: one where the force increases and decreases, and a second where the force reaches more or less constant values, i.e. static and kinetic, respectively. Although the transition from static to kinetic frictional force may seem abrupt, it is actually continuous. The coefficients of friction can be calculated based on $F = \mu_s N$ and $F = \mu_k N$, where the frictional force is F, N is the magnitude of the normal force and μ_s and μ_k are the coefficients of static and kinetic friction, respectively. For a predetermined surface, such coefficients are reasonably constant, and theoretically independent of the contact area and, in the case of kinetic force, the speed of the relative motion (Halliday & Resnick, 1988). Both static and kinetic coefficients increased with the roughness of the screened gel, i.e. the smooth surface had the lowest static and kinetic friction coefficients and the surface with maximal roughness yielded the highest coefficients, whereas the surfaces with mid-roughness yielded intermediate coefficients (Ben-Zion & Nussinovitch, 1997b).

Figure 1.11 exhibits the positive trend between achieved peel-bond strength and the respective coefficients of friction, μ_s and μ_k. Each point in this figure represents the average of five experiments, and five degrees of roughness. Strong linear relationships were calculated for both trends (r = 0.92 and 0.98 for the static and kinetic coefficients, respectively). The identical trends for both frictions verify the relationships between friction and adhesion, and justify the choice of the friction test to characterize the property of a potential gel for adherence. Figure 1.12 represents the tensile-bond strength obtained from three types of agar, each poured as a hot solution onto its respective bottom layer. It should be noted that this is the only test which can be used to measure bond strength since agars are generally brittle, except for *Gelidium japonicum* extract. From Fig. 1.12, it is clear that at least for the *G. japonicum* extract and for the agarose, reducing the temperature at which the upper hot solution is

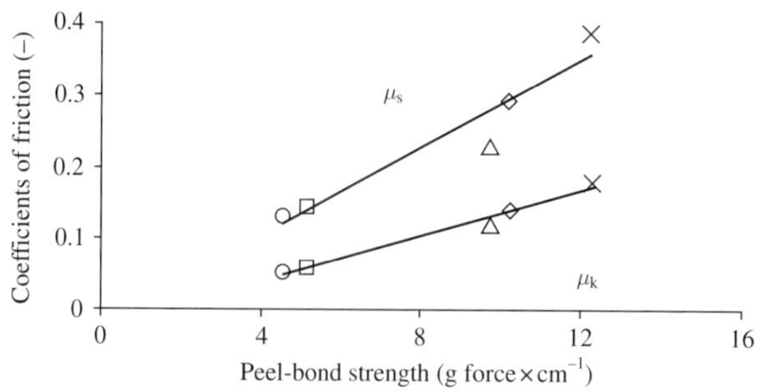

Fig. 1.11 The correlation between coefficients of friction and peel-bond strength obtained for five roughnesses used in this work. o, smooth surface; □, width = 1.00 mm, depth = 0.56 mm; △, w = 0.5 mm, d = 0.32 mm; ◇, w = 0.2, d = 0.14 mm; ×, w = 0.16, d = 1.112 mm. (From Ben-Zion & Nussinovitch, 1997c, with permission from Elsevier Science.)

poured results in a reduction in the bond strength between vicinal layers. A top-layer solution whose temperature is equal to or above the melting point ensures even melting of the mono-layer, and hence a well-bonded layer assembly (Ben-Zion & Nussinovitch, 1997a). In general terms, the kind of agar used to construct the gel layers significantly affects the adhesion pattern. Factors deriving from the molecular helical shape, degree of substitution and distribution of the chains may play a role in the inter-molecular interactions. In addition, the higher the volume of a particular solution, the better it maintains its high temperature. No significant differences were found between the layers of different thicknesses in terms of their peel-bond strength. It was suggested that if the heat capacity of the poured layer is large enough to provide melting at the interface, resulting in molecular interactions, an increase in volume will not elevate the measured strength (Ben-Zion & Nussinovitch, 1997b).

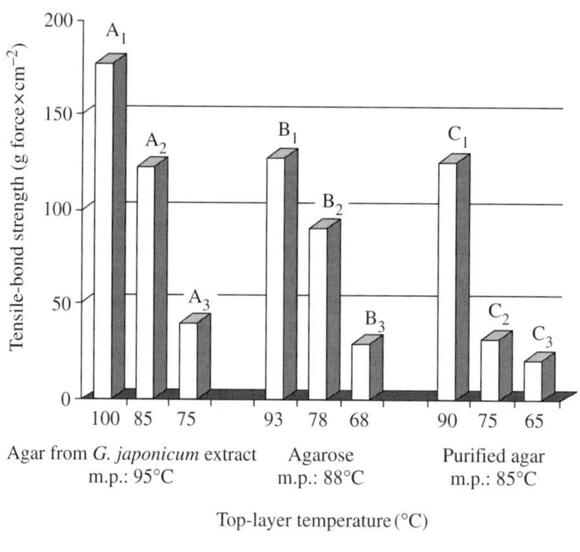

Fig. 1.12 Tensile-bond strength for double-layered gels made with three different agar solutions cooled to various temperatures before pouring onto the identical bottom-layer gel. (From Ben-Zion & Nussinovitch, 1997c, with permission from Elsevier Science.)

1.4.4 Hydrocolloids as wet glues

Twenty-six hydrocolloids were studied for their ability to create very thick suspensions with *good* adherence properties at predetermined gum loadings ranging from 10 to 75% (w/w). The hydrocolloids were: gum talha, gum ghatti, AG, gum karaya, gum tragacanth, dextran, apple pectin, CMC, HPMC, tapioca-dextrin, carbopol, HPC, MC, gelatin, casein, starch, LBG, guar gum, alginate, κ-carrageenan, tara gum, fenugreek gum, konjak mannan, xanthan gum, gellan and curdlan. The hydrocolloids (at different concentrations) were added in powdered form to double-distilled water and mixed to obtain thick, uniform and smooth wet-paste glues (Ben-Zion & Nussinovitch, 1997a).

Preliminary tests revealed that only 13 of the hydrocolloids, namely gum talha, gum ghatti, gum karaya, gum tragacanth, AG, dextran, pectin, tapioca-dextrin, CMC, MC, HPC, HPMC and carbopol, can serve as bioadhesives in hydrophilic systems. Table 1.1 gives some of their physical properties. All preparations were tested immediately following their production. The pHs of the wet glues ranged from 1.2 in carbopol to ~9.6 in AG. A variety of different colors could be found among the wet glues. Wet glues with L^* values of ~60 were found to be the lightest of all the glues tested (the higher the L^* value, the lighter the paste). People tend to prefer pastes with a 'whitish' appearance. The a^* value spans the green–red axis of the color system and the b^* value, the yellow–blue axis. Many of the wet glues had a yellowish appearance (higher b^* values). Only a few were dark brown (Ben-Zion & Nussinovitch, 1997a).

Table 1.2 shows the results of the 90°-peel test for the 13 hydrocolloids at various concentrations. This is a common method for testing pressure-sensitive adhesives (ASTM D-2860), rubber cements (ASTM D-816), fabric-adhering rubber compounds (ASTM D-4393), metallic coatings (ASTM B-571) and the like. The gums karaya, tragacanth, ghatti and talha, tapioca-dextrin, AG and dextran all behaved like sticky adhesive substances. This phenomenon was observed at concentrations from 45 to 75% (w/w). Pectin, carbopol-934 and cellulose derivatives began to behave like wet adhesives at concentrations ranging from 15 to 35% (w/w). For almost all individually tested gums, the higher the concentration of the hydrocolloid within the wet glue, the higher was its peel-bond strength. Pectin and MC showed a clearly different behavior, whereby the strength decreased as the percentage of hydrocolloid within the glue increased. This is an inherent material property and is caused by the different water-holding capacities of the networks. All peel-bond strength values measured in the system, using SSM (a surface that simulates human skin and is composed primarily of protein, lipids and water), were smaller than those found for the cellulose–acetate system. The exception was carbopol, where no significant differences in strength with either substrate were observed (Table 1.2). Thus, the measured estimation of failure was controlled by the cohesiveness of the carbopol wet glue, and separation was observed within the glue layer. Since the amount of water in the SSM was ~25%, as compared to less than 3% in the cellulose–acetate membrane, the affinity of the hydrocolloid adhesives to the latter was higher (Ben-Zion & Nussinovitch, 1997a).

No information on the wetting of substrates is available, as hydrocolloid wet glues are in the form of a very thick paste and can only be smeared on film with applied stress manipulation. No contact angles can be measured since no drops of such glues

Table 1.1 Physical properties of hydrocolloid pastes.

Hydrocolloid	Loading (g/100 g)	pH[a]	Color parameters *L**	*a**	*b**	Hue
Tree and shrub exudates						
Gum talha	65	4.02	22.16	0.00	0.43	Dark brown
	70	3.98	21.69	0.59	1.08	"
	75	3.95	24.21	3.07	4.03	"
Gum karaya	25	3.68	27.98	2.00	4.95	Light brown
	35	3.61	29.17	2.09	4.82	"
	45	3.52	26.94	2.51	5.40	"
Gum tragacanth	25	4.65	43.94	–0.92	6.26	Yellowish
	35	4.61	47.76	–0.34	8.73	"
	45	4.56	52.51	–0.11	11.67	"
Gum ghatti	55	4.22	23.65	1.18	2.55	Dark brown
	65	4.17	23.69	1.45	2.64	"
	75	4.12	24.82	1.68	2.58	"
Tree extracts						
AG	65	9.57	69.08	–2.35	12.45	Yellow
	75	9.08	60.13	–0.29	25.19	"
Fruit extracts						
Apple pectin	15	2.32	62.70	–1.5	6.84	Yellowish-white
	25	2.28	64.55	–0.72	9.77	"
Grains						
Tapioca	55	3.11	62.68	–0.11	8.76	Yellowish
	65	2.98	65.34	0.67	11.20	"
Exocellular polysaccharides						
Dextran	65	4.72	46.09	–0.29	1.05	White
	70	4.89	61.80	–0.14	0.54	"
Cellulose derivatives						
CMC	10	6.87	52.77	–0.90	3.64	Yellowish transparent
	15	7.06	34.05	–0.49	5.73	"
	20	7.08	35.05	–0.25	5.45	"
	25	7.12	39.57	–0.13	6.24	"
MC	10	5.18	32.86	–0.34	0.81	Light white
	15	5.11	45.73	–0.65	1.58	"
	20	5.07	54.78	–0.69	2.07	"
	25	4.85	59.29	–0.75	2.70	"
HMPC	10	4.49	31.25	–0.58	2.02	Semi-transparent
	15	4.48	29.17	–0.44	1.74	"
	20	4.42	31.22	–0.43	1.99	"
	25	4.37	34.38	–0.42	2.26	"
HPC	10	4.10	30.30	–0.52	1.40	"
	15	4.08	29.71	–0.40	1.41	"
	20	4.02	31.46	–0.44	1.35	"
	25	3.99	31.08	–0.51	1.83	"
Petrochemicals						
Carbopol-934	15	1.90	36.05	–0.60	–0.39	White
	25	1.42	32.88	–0.55	0.83	"
	35	1.24	38.48	–0.87	3.19	"

[a]pH of water = 6.5.

Table 1.2 Peel-bond-strength values for 13 hydrocolloid pastes, tested on cellulose-acetate film and on SSM.

Hydrocolloid	Loading (g/100 g)	Peel-bond strength (g force/cm), Means (SD): Cellulose-acetate membrane	SSM*
Tree and shrub exudates			
Gum talha	65	5.9 (1.1)	3.6 (0.4)
	70	18.8 (1.6)	7.5 (0.6)
	75	45.2 (2.5)	15.4 (1.3)
Gum karaya	25	4.9 (0.7)	3.4 (0.5)
	35	12.2 (0.7)	10.7 (0.8)
	45	31.4 (2.4)	26.3 (2.1)
Gum tragacanth	25	4.4 (0.6)	3.4 (0.4)
	35	11.9 (0.6)	9.4 (0.3)
	45	30.9 (1.1)	23.2 (0.9)
Gum ghatti	55	7.1 (0.9)	6.0 (1.0)
	65	22.0 (1.6)	11.5 (0.6)
	75	62.6 (1.4)	0.0 (0.0)
Tree extracts			
AG	65	4.2 (0.6)	2.6 (0.4)
	75	14.3 (2.2)	6.4 (0.9)
Fruit extracts			
Apple pectin	15	9.3 (0.5)	6.7 (1.0)
	25	6.0 (0.6)	4.0 (0.6)
Grains			
Tapioca-dextrin	55	8.5 (0.8)	5.2 (1.3)
	65	16.8 (1.2)	14.7 (1.9)
Exocellular polysaccharides			
Dextran	65	7.4 (0.9)	4.5 (0.5)
	70	62.5 (1.9)	49.6 (1.8)
Cellulose derivatives			
CMC	10	2.5 (0.3)	1.6 (0.4)
	15	3.3 (0.4)	2.3 (0.5)
	20	6.3 (0.7)	2.7 (0.2)
	25	4.8 (0.6)	2.6 (0.4)
MC	10	9.2 (0.6)	9.4 (0.6)
	15	7.3 (0.7)	7.2 (0.5)
	20	6.0 (0.4)	4.9 (0.5)
	25	3.8 (0.5)	1.8 (0.3)
HMPC	10	2.7 (0.5)	1.4 (0.1)
	15	4.8 (0.4)	3.0 (0.7)
	20	9.1 (0.7)	6.3 (1.0)
	25	7.0 (0.6)	5.9 (0.5)
HPC	10	1.7 (0.4)	2.0 (0.5)
	15	2.5 (0.4)	3.3 (0.3)
	20	3.9 (0.5)	3.9 (0.5)
	25	7.7 (0.4)	5.0 (0.7)
Peterochemicals			
Carbopol-934	15	4.1 (0.8)	4.1 (1.1)
	25	13.7 (1.0)	14.8 (1.5)
	35	38.8 (1.3)	40.4 (2.7)

*Water content = 25%.

can be formed and flow cannot be induced by regular means. Typical results for the 90°-peel, tensile-load and lap-shear tests are presented in Fig. 1.13. Two common types of curves were obtained from the peel test. When a sample is pulled apart at a constant cross-head speed, the measured force should ideally be constant after reaching a steady-state condition. In practice, this is not always the case (Fig. 1.13a). When such results are reported, the mean values (averages) of a curve's ruggedness (deviation from a smooth line after reaching a steady-state condition) can be calculated and observed. In some instances, when specimens are observed macroscopically at the time of testing, the rupture process occurs abruptly, sample failure propagating faster than the rate at which the sample is being pulled apart, and failure is initiated periodically. The force has been claimed to go through a well-defined maxima and minima, and the distance between two minima or maxima to be independent of testing rate (Ben-Zion & Nussinovitch, 1997a). Figure 1.13a shows that certain systems exhibit *slip-stick* failure, whereas the failure process is continuous in others (Fig. 1.13b). The latter case shows a steady-state force when failure is initiated continuously. Tensile-load (Fig. 1.13c) and lap-shear (Fig. 1.13d) tests can be applied and results plotted as load ($g\,cm^{-2}$) versus displacement (cm). The tensile-bond strength increases in parallel to increases in deformation until the beginning of failure. Tensile-bond tests are commonly used to test various adhesives, ranging from those for wood to those for metal (ASTM D-897 and ASTM D-2094, respectively). The lap-shear strength decreases linearly as deformation increases. Lap-shear tests are used to examine adhesion when the samples are relatively easy to construct and when they closely resemble the geometry of the many practical joints (ASTM D-1002 and ASTM D-3528). Table 1.3 gives the experimental values of tensile-bond and lap-shear strengths for seven hydrocolloids, representing seven different origins of natural and synthetic gums. Under the test conditions, carbopol gave the highest tensile-bond strength, dextran the lowest. It is important to note that the concentrations studied were not the same, since the concentrations to produce the highest bond strength in each gum were chosen based on the previously performed peel test. Dextran differed from the other gums in its ability to dry rapidly after preparation, making it difficult to use at higher concentrations. In the lap-shear test, tapioca was found to produce the highest lap-shear strength, dextran the lowest. This test was performed to examine the cohesiveness of the glue, because the adhesive pressed between two sliding glass plates undergoes shearing in the medium itself, and not at the interface between the medium and the adhered substrate (Ben-Zion & Nussinovitch, 1997a).

The increase in adhesive strength paralleled that of the deformation rate, suggesting that the adhesive bond is of a viscoelastic nature, i.e. that at faster rates of stress application the adhesive bond has less time to deform and flow. At higher cross-head speeds, the peel-bond strength tends to be rate-independent because the interactive forces at the interface are characterized by their short range (Ben-Zion & Nussinovitch, 1997a). To study the influence of water absorbance on peel-bond strength values, SSM films were produced with different moisture contents by immersing them in water for 2, 5, 9 and 13 min. Thus, films were produced with water contents of ~31, 39, 46 and 56%, respectively. AG, dextran, HPMC, gum ghatti and tapioca-dextrin exhibited decreases in peel-bond strength as the water content increased. Hydrocolloids develop

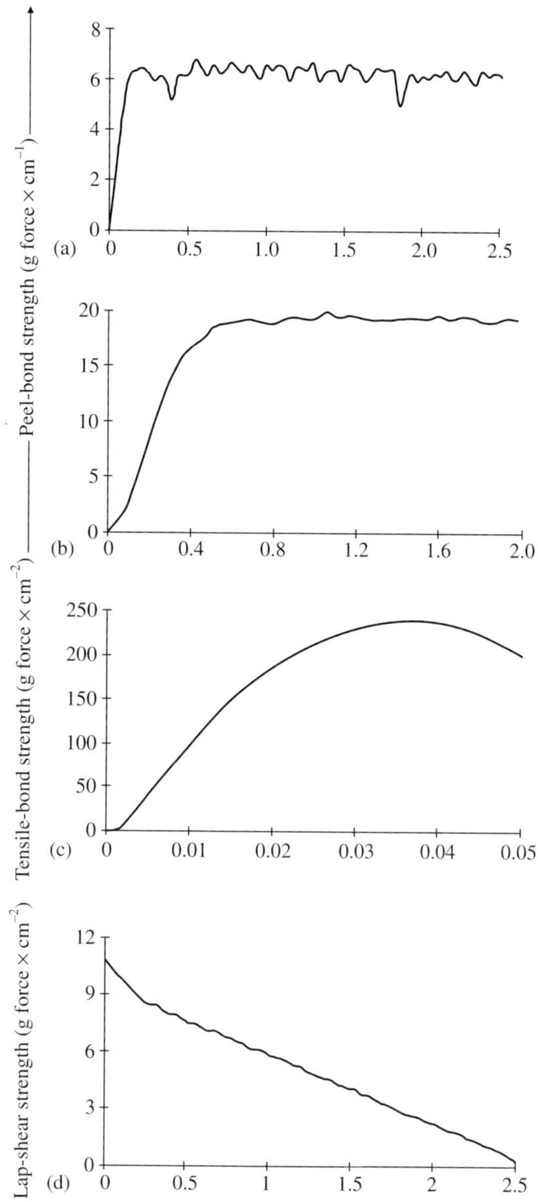

Fig. 1.13 Typical curves for the 90°-peel test. (a) pectin, 25%; (b) gum talha, 70%; (c) typical curve for the tensile-bond test (dextran, 65%); (d) typical curve for the lap-shear test (HPMC, 20%). (From Ben-Zion & Nussinovitch, 1997a, with permission from Elsevier Science.)

their wet adhesive properties at various degrees of hydration, reaching maximum adhesion at an optimum degree of hydration. Some hydrocolloids exhibit wet adhesiveness in the presence of only very little water, whereas excessive amounts of water cause the formation of a slippery, non-adhesive mucilage. A wet surface is not static and water diffuses from the surface of the wet substrate (SSM) into the interface of the

Table 1.3 Tensile-bond and lap-shear test values for bond strength of seven hydrocolloid adhesive pastes.

Hydrocolloid	Concentration (g/100 g)	Tensile-bond strength (g force/cm^2)	COV*	Lap-shear strength (g force/cm^2)	COV*
Dextran	65	232.8	2.0	8.1	13.8
Carbopol-934	35	1112.7	1.5	46.1	2.3
Ghatti	65	452.8	4.9	27.1	1.3
Tapioca-dextran	65	884.6	2.2	49.7	5.7
AG	75	568.7	2.3	29.5	2.7
Pectin	15	334.7	11.5	20.4	1.3
HPMC	20	259.2	5.8	12.4	17.4

*COV is the coefficient of variance.

hydrocolloid. The rate and capacity of water absorbance of the hydrocolloids affect the amount of water present near the interface between the adhesive and the substrate; thus, the water molecules serve as plasticizing agents. These properties appear to be important in determining the time required for initial wet adhesion, hydration time and duration of adhesion. Rapid water absorbance may shorten the duration of adhesion because erosion proceeds rapidly. An excessive amount of water at the interface causes over-extension of the hydrogen bonds and other adhesive forces, leading to a weakening of adhesive-bond strength. The mechanism of wet adhesion is based on the following concept and explanation by Chen and Cyr (1970). They state that when a dry hydrocolloid is hydrated, the long chains of the polymer are liberated to a freely moving state. The stretched, entangled or twisted molecules are able to match their active adhesive sites with those on the hydrophilic substrate to form adhesive interactions, or match with each other to form cohesive bonds. When an optimum amount of water is present, at or near the interface, perfect matching is attained. A hydrocolloid that is hydrated by insufficient amounts of water will not possess completely liberated and exposed adhesive sites. In the case of carbopol-934 and pectin, no significant change in peel-bond strength was found when more than ~50% water was present in the SSM. This suggests that adhesion is not affected by such degrees of hydration and that the constant peel-bond strength is a result of the cohesive force being measured (Ben-Zion & Nussinovitch, 1997a).

Figure 1.14 represents the relationship between peel-bond strength and water number (amount of water evaporated from a standard sample) and the drying time of the wet glues produced from the seven gums representing the various natural and synthetic hydrocolloids in this study. The wet glue served as an adhesive between the covering cellulose acetate membrane and the SSM layer with ~1% water content. The hydrocolloids were found to behave in different ways. The peel-bond strength increased with time (up to ~4 h for dextran and 8 h for gum ghatti and AG) and then decreased. This is a result of the gradual drying of the wet glues. For the tapioca-dextrin, peel-bond strength increased with time up to 1 h, and then decreased gradually. For carbopol, pectin and HPMC, a gradual increase in peel-bond strength was observed over the whole test period. For pectin and HPMC, adhesion strength values were smaller than those of the other adhesives. After 8 to 12 h, peel-bond strength for all hydrocolloid glues, except carbopol, pectin and HPMC, decreased to a value of ~0. This is due to the fact that the wet glue dries and adheres strongly to the SSM only, while no attachment

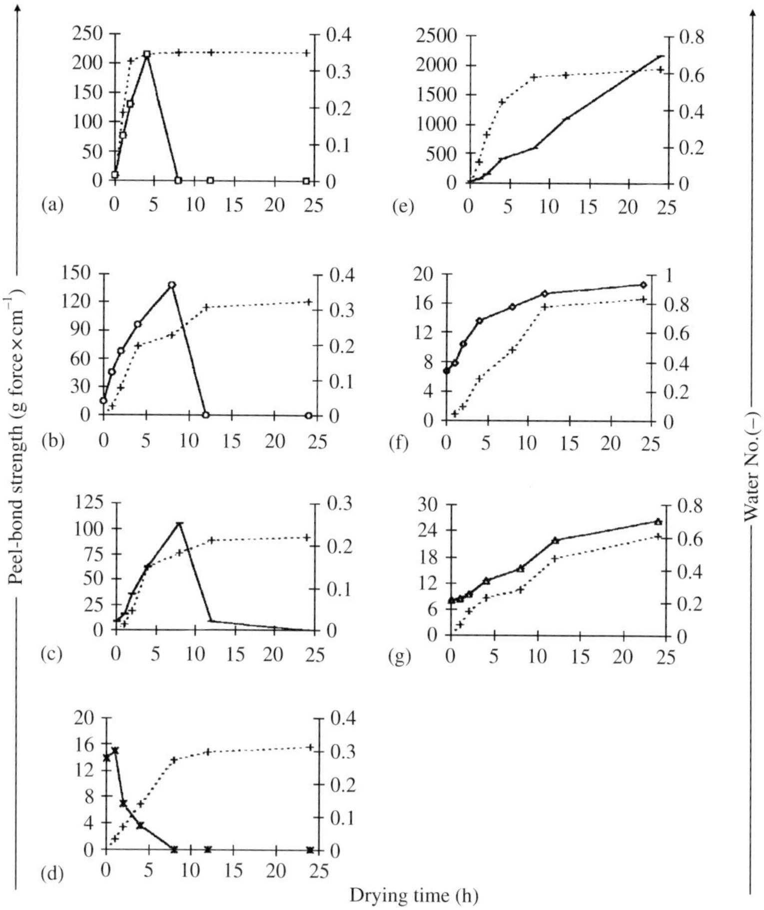

Fig. 1.14 Relationships between contact time of the covering cellulose acetate film adhered to SSM substrate with 1% water content by hydrocolloid adhesive and 90°-peel-bond strength affected by water loss from the paste expressed as water number. The dashed lines represent the water number; the solid line represents the peel-bond strength. Gum concentrations given as (g per 100 g): (a) dextran (65), (b) ghatti (65), (c) AG (75), (d) tapioca (65), (e) carbopol (35), (f) pectin (15), (g) HPMC (20). (From Ben-Zion & Nussinovitch, 1997a, with permission from Elsevier Science.)

exists with the cellulose acetate film. Failure occurs at the interface of the glue–substrate array, thus, exhibiting adhesion-type mode. The water number for all tested hydrocolloids increased with drying time, except in the case of HPMC and pectin, where the initial water content of the glue was higher than in the other systems.

1.5 Other uses and future prospects

Other uses of adhesives in the food arena include: greeting cards made from a cardboard base decorated with sweets, which are stuck onto the card with an edible adhesive made of sugar, gelatin and water (Petrovic, 1998); a seasoning for frozen foods comprising a dry mixture of seasoning salt and fat, which contains 5–80% by weight, preferably

5–50%, of an adhesion-improving agent from a native starch, yeast powder, soy powder, maltodextrin and skim-milk powder (Ammedick Naumann & Caroly, 1997).

In around 1930, water-based glues made up ~90% of the adhesives market. From 1930 to 1986, this figure decreased to ~60%. However, an average annual increase of ~3.9% in the last 10 years in these glue types, due to renewed interest from the packaging, construction and medicinal industries, suggests that nature-based products will still make up 48% of the market in the new millennium, whereas early synthetic adhesives, including PVA, PVP and polyacrylic acid, will take 16% of the market share and new synthetic adhesives ~36%.

1.6 References

Ammedick Naumann, C. & J. Caroly (1997) Seasoning mixture for the frozen foods industry. German Federal Republic Patent Application DE19535582A1.

Anon (1979) Saturation of crumb coatings with oil without frying. *Research Disclosure No. 182, 322.* Alginate Industries Ltd. London, UK.

Anon (1993) New adhesive improves quality of baked snacks. *Food Engineering*, **65**(4), 34.

Bauman, M.G.D. & A.H. Conner (1994) Carbohydrate polymers as Adhesives. In: (Pizzi, A., Ed.), *Handbook of Adhesive Technology*. Marcel Dekker Inc., NY, pp. 299–313.

Ben-Zion, O. & A. Nussinovitch (1996) Predicting the deformability modulus of multi-layered texturized fruits and gels. *Lebensm.-Wiss. u.-Technol.*, **29**, 129–134.

Ben-Zion, O. & A. Nussinovitch (1997a) Physical properties of hydrocolloid wet glues. *Food Hydrocolloids*, **11**(4), 429–442.

Ben-Zion, O. & A. Nussinovitch (1997b) A prediction of the compressive deformabilities of multilayered gels and texturized fruit, glued together by three different adhesion techniques. *Food Hydrocolloids*, **11**(3), 253–260.

Ben-Zion, O. & A. Nussinovitch (1997c) Adhesion-strength characteristics of double-layered agar-galactomannan mixed gels. *Food Hydrocolloids*, **11**(4), 373–384.

Bottenberg, P., R. Cleymaet, C. de-Muynck, J.P. Remon, D. Coomans, Y. Michotte & D. Slop (1991) Development and testing of bioadhesive fluoride containing slow-release tablets for oral use. *J. Pharm. Pharmacol.*, **43**, 457–464.

Bouckaert, S. & J.P. Remon (1993) In-vitro bioadhesion of a buccal, miconazole slow-release tablet. *J. Pharm. Pharmacol.*, **45**, 504–507.

Brief, A. (1990) The role of adhesives in the economy. In: (Skeist, I., Ed.), *Handbook of Adhesives*. Van Nostrand Reinhold, NY, pp. 21–38.

Campbell, C.S. (1972) Battermix starches. United States Patent 3,655,443.

Chen, J. & G.N. Cyr (1970) Compositions producing adhesion through hydration. In: (Manly, R.S., Ed.), *Adhesion in Biological Systems*. Academic Press, NY, pp. 163–181.

Croll, S.G. (1983) Adhesion and internal strain in polymeric coatings. In: (Mittal, K.L., Ed.), *Adhesion Aspects of Polymeric Coatings*. Plenum Press, NY and London, pp. 107–129.

Dea, I.C.M. (1979) Interaction of ordered polysaccharides structure-synergism and freeze-thaw phenomena. In: (Blanshard, J.M.V. & J.R. Mitchell, Eds.), *Polysaccharides in Food*. Butterworth, London, p. 238.

Every, D. & M. Ross (1996) The role of dextrins in the stickiness of bread crumb made from pre-harvest sprouted wheat or flour containing exogenous alpha-amylase. *J. Cereal Sci.*, **23**(3), 247–256.

Fellows, P.J. (1990) *Food Processing Technology*. Ellis Horwood, NY.

Gates, J.G. (1981) *Basic Foods*. Holt, Rinehart and Winston, NY.

Glicksman, M. (1982) *Food Hydrocolloids*. CRC Press Inc., Boca Raton, FL, **3**, p. 176.

Hagqist, J., K.F. Meyer & K.M. Sandra (1990) Adhesives market and applications. In: (Dostal, C.A., Ed.), *Adhesives and Sealants*. ASM International, London, pp. 87–88.

Halliday, D. & R. Resnick (1988) *Fundamentals of Physics*. John Wiley & Sons, NY, pp. 104–107.

Harada, T. (1977) Production properties and application of curdlan (with applications to the food industry). In: (Sanford, A., Ed.), *Extracellular Microbial Polysaccharides*. ASC Symposium Series, Washington, DC, pp. 265–283.

Harada, T. (1979) Curdlan: a gel-forming β-1,3 glucan. In: (Blanshard, J.M.V & J.R. Mitchell, Eds.), *Polysaccharides in Food*. Butterworths, London, pp. 283–300.

Hartshorn, S.R. (1986) In: (Hartshorn, S.R., Ed.), *Structural Adhesives Chemistry and Technology*. Plenum Press, NY and London, pp. 1–19.

Howes, F.N. (1949) In: *Vegetable Gums and Resins*. Chronica Botanica Comp., Waltham, MA, pp. 56–58, 61–62.

Hsia, H.Y., D.M. Smith & J.F. Steffe (1992) Rheological properties and adhesion characteristics of flour-based batters for chicken nuggets as affected by three hydrocolloids. *J. Food Sci.*, **57**(1), 16–18, 24.

Ikeda, T., S. Moritaka, S. Sugiura & T. Umeki (1976) Method for preparing jelly foods. United States Patent 3,969,536.

Irons, B.K. & J.R. Robinson (1994) Bioadhesives in drug delivery. In: (Pizzi, A., Ed.), *Handbook of Adhesive Technology*. Marcel Dekker Inc., NY, pp. 615–627.

Kadison, S. & D. Scheiner (1985) Gelatin dessert product. United States Patent 4,500,552.

Kanig, J.L. & P. Manago-Ulgado (1965) The in-vitro evaluation of orolingual adhesives. *J. Oral Therap. Pharm.*, **4**, 413–420.

Keusch, P. & J.L. Essmyer (1987) Adhesive polyethylene oxide hydrogel sheet and its production. United States Patent 4,684,558.

Kiyosi, O. & S. Yasuo (1986) Pullulan-containing adhesive tapes, sheets and labels. *Chem. Abstr.*, 106.

Kuntz, L.A. (1997) Catching value in seafood. *Food Product Design*, **7**(9), 71, 73–74, 77–78, 81–82.

LaBell, F. (1994) Coating systems for shrimp. *Prepared Foods*, **163**(3), 59–60.

McGlinchey, N. (1994) Batter by far speciality starches in battered and breaded foods. *Food Tech. Europe*, **1**(5), 96, 98, 100.

Mantell, C.L. (1947) In: *Water-Soluble Gums*. Reinhold Pub. Corp., NY, pp. 48, 71, 72.

May, K.N., A.J. Farr & J.P. Hudspeth (1969) Estimating breading content of battered and breaded poultry parts. *Food Technol.*, **23**, 1087–1090.

Mazurkiewicz, J., H. Zaleska & J. Zaplotny (1993) Studies in carbohydrate-based glues and thickeners for foodstuffs. I. Glucose-sucrose-apple pectin ternary system. *Starch*, **45**(5), 175–177.

Meyers, A.M. (1990) Functionality of hydrocolloids in batter coating system. In: (Kulp, K. & R. Loewe, Eds.), *Batters and Breadings in Food Processing*. American Association of Cereal Chemists, Inc., St. Paul, MN, USA, pp. 117–142.

Nippon Suisan Kaisha Ltd. (1973) Frozen food coatings. Japanese Patent 4,834,902.

Nussinovitch, A. (1997) In: *Hydrocolloid Applications: Gum Technology in the Food and Other Industries*. Blackie Academic & Professional, London, pp. 14–15.

Petrovic, S. (1998) Greeting card, message, etc. made with sweets. French Patent Application FR2759021A.

Piglowski, J. & M. Kozlowski (1985) Rheological properties of pressure-sensitive adhesives: polyisobutylene/sodium carboxymethylcellulose. *Rheol. Acta*, **24**, 519–524.

Portelli, G.B. (1986) In: (Hartshorn, S.R., Ed.), *Structural Adhesives Chemistry and Technology*. Plenum Press, NY and London, pp. 407–449.

Quigg, J.R. (1980) Increased consumption of varied seafoods demands new batters and breadings. *Quick Frozen Foods*, **43**(9), 28–30.

Rispoli, J.M., H.H. Sergenian, H. Topalian, M.A. Rogers & J.S. Swartley (1987) Coating mix containing a fried component and process therefore. European Patent EP0110587B1.

Robinson, J.R., M.A. Longer & M. Veillard (1987) Bioadhesive polymers for controlled drug delivery. In: (Juliano, R.L., Ed.), *Controlled Delivery of Drugs*. Acad. Sci., Ann Arbor, MI, pp. 507, 307–314.

Sadle, A., N.J. Norristown & T.J. Pratt (1979) Starch carrier composition for adhesive containing urea as a selatinizing agent. United States Patent 4,157,318.

Shay, K. (1991) Denture adhesives – choosing the right powders and pastes. *J. Amer. Dent. Assoc.*, **122**, 70–76.

Smart, J.D. (1991) An in-vitro assessment of some mucosa dosage forms. *Int. J. Pharm.*, **73**, 69–74.

Smart, J.D., I.W. Kellaway & H.E.C. Orthington (1984) An in-vitro investigation of mucosa-adhesive materials for use in controlled drug delivery. *J. Pharm. Pharmacol.*, **36**, 295–299.

Suderman, D.R. (1980) Factors affecting the adhesion of coating to poultry skin. *Dissertation Abstracts International*, **41**(1), 122–123.

Suderman, D.R. (1992) Application of batters and breadings to poultry, seafood, red meat and vegetables. In: (Kulp, K. & R. Loewe, Eds.), *Batters and Breadings in Food Processing*. American Association of Cereal Chemists, Inc., St. Paul, MN, USA, pp. 177–198.

Suderman, D.R. & F.E. Cunningham (1979) New portable sieve shaker tags breading adhesion. *Broiler Ind.*, **42**, 66–67.

Suderman, D.R., J. Wiker & F.E. Cunningham (1981) Factors affecting adhesion of coating to poultry skin: effects of various protein and gum sources in the coating composition. *J. Food Science*, **46**(4), 1010–1011.

Sutherland, J.P., A.H. Varnam & M.G. Evans (1986) *A Color Atlas of Food Quality Control*. A Wolf Science Book, Weert, The Netherlands.

Toloday, D. & C. Andres (1975) Vegetable gum dry mix improves shrimp breading. *Food Processing*, **36**(5), 32.

Torrey, S. (1977) In: *Adhesive Technology Developments*. Noyes Data Corporation, NJ, pp. 197–200.

Toulmin, H.A. (1956) United States Patent 2,749,277.

Valli, R., B. Chinn & B. Heidolph (1997) Sweet doughs laminated with hydrocolloid solutions. *NutraSweet Kelco Co. Research Disclosure No. 398, 418.*

Chapter 2
Hydrocolloid Coatings

2.1 Introduction

Hydrocolloids (gums) are high-molecular-weight molecules, usually with colloidal properties, which, in an appropriate solvent, produce solutions or highly viscous suspensions or gels with low dry substance content. The term gum is applied to a wide variety of substances with *gummy* characteristics. Most commonly, the term gum in the technical sense in industry refers to plant or microbial polysaccharides or their derivatives that are dispersed in either cold or hot water to produce viscous mixtures or solutions.

Hydrocolloids are used to produce thin layers of edible materials on food surfaces or between food components. Such films serve as inhibitors of moisture, gas, aroma and lipid migration. They can include antioxidants, antimicrobial agents (Kester & Fennema, 1986), preservatives or other additives to improve mechanical integrity or handling characteristics and food quality, and to change surface gloss (Nussinovitch, 1998). Many gums and their derivatives have been used for coating purposes. They include alginate, carrageenan, cellulose and its derivatives, pectin, starch and its derivatives, among others. Since these gums are hydrophilic, the coatings they produce have nature-limited moisture-barrier properties. However, if they are used in a gel form (with no drying), they can retard moisture loss, for example from meats and soft white-brined cheeses, during short-term storage when the gel acts as a sacrificing agent rather than a barrier to moisture transmission (Kampf & Nussinovitch, 1998, 2000; Kester & Fennema, 1986). In addition, since in some cases an inverse relationship between water vapor and oxygen permeability has been observed, such films can provide effective protection against the oxidation of lipids and other susceptible food ingredients.

Most of the available information on food coatings focuses on moieties created by drying hydrocolloid solutions or blends. This chapter emphasizes, but by no means limits its discussion to, coatings created by drying hydrocolloid gels and wax–hydrocolloid mixtures. In addition, gloss has become a forgotten food property, and the way in which food surface gloss can be changed by coating films is discussed in only a few manuscripts; thus this topic is discussed in detail in this chapter.

2.2 Edible packaging materials – a general approach

Reviews on food-packaging materials based on natural polymers can be found elsewhere (Cuq *et al.*, 1997; Debeaufort *et al.*, 1998; McHugh, 1996; Nussinovitch, 1997, 1998, 2000; Nussinovitch & Lurie, 1995). The three main categories of macromolecules found in edible films are polysaccharides, proteins and lipids (McHugh, 1996). Polysaccharide and protein films are good gas barriers, but poor moisture barriers. Conversely, pure lipid films are good moisture barriers, but poor gas barriers. Considerable attention has been focused on the development of composite edible films that utilize the favorable functional characteristics of each class of macromolecules (McHugh, 1996). Research into the nature of polysaccharide–lipid edible films has therefore generated much interest, and resultant films exhibit common bilayer structures with drastically improved impermeability. These and similar films will be discussed later in this chapter.

Food-packaging materials include biodegradable packaging materials, and those based on mixtures of synthetic polymers and biopolymers. The main types of biopolymers of possible use in packaging materials are starch, cellulose, pullulan, xanthan, gellan, chitosan, proteins and polymers of hydroxylated fatty acids, among others (Cuq *et al.*, 1997).

Edible coatings should be tailored to suit the food product they are designed to enrobe. Organoleptic properties of coatings (usually tasteless) are very important for consumable food products. Another option is to deliberately change the taste of the coating, allowing its sensorial properties to create a nice blend with those of the product itself. Besides being tasty or tasteless, the coatings should have the required mechanical properties and tailor-made barrier stability. Other relevant benefits are simple application, biodegradability and non-toxicity, safety for the user and consumer, and lesser cost (Debeaufort *et al.*, 1998).

2.3 Uses of hydrocolloids for coating

2.3.1 Past and present gum coatings for fruits and vegetables

Vegetables add color, flavor and texture to our diets. They are also good sources of many nutrients that are not plentiful in other foods. The fruit–vegetable distinction is determined mostly by custom. Botanically, a fruit is the seed-bearing part of a plant, which means that squashes, cucumbers, green beans, peas in the pod, okras and tomatoes are actually fruits, even though these plants are usually regarded as vegetables. Technological developments in recent years have made a wide variety of fruits more plentifully available than ever before (Gates, 1981).

An acceptable estimate is that 25%, and up to 80%, of freshly harvested fruits and vegetables are lost through spoilage (Willis *et al.*, 1981). Fruits and vegetables continue their metabolic activity after harvest and will either ripen if climacteric, or senesce if non-climacteric rather rapidly unless special procedures are adopted to slow down these processes (Nussinovitch, 1997). Post-harvest shelf-life extension can be achieved by the application of edible coatings, which are semi-permeable to water vapor and gases. Such coatings can enhance or replace other techniques used for the same purpose such as modified atmosphere (MA) or controlled atmosphere storage (Barkai-Golan, 1990; Smith *et al.*, 1987). Other achievements of coating applications include improvement of mechanical handling properties (Mellenthin *et al.*, 1982), and retention of volatile flavor compounds (Nisperos-Carriedo *et al.*, 1990).

The use of wax coatings (Fig. 2.1) for citrus fruits can be traced back to the twelfth and thirteenth centuries (Hardenburg, 1967). In the 1930s, paraffin-based waxes were used to coat citrus fruits and vegetables, and in the 1950s, carnauba wax–oil emulsions were used (Kaplan, 1986). Different ingredients, such as fungicides, chilling-injury protectants, coloring agents, anti-senescence substances, growth regulators and biological control agents can be incorporated within the coatings (Baldwin, 1994; Brown, 1984; McQuilken *et al.*, 1992; Miller *et al.*, 1988; Radnia & Eckert, 1988). Coatings based on acetylated monoglycerides (AGs), waxes and surfactants were developed to eliminate problems associated with apples, such as moisture loss, surface abrasion, soft scald, coreflush and spots (Fisher & Britton, 1940; Hardenburg, 1967; Kester & Fennema, 1988; Lawrence & Iyengar, 1983; Paredes-Lopez *et al.*, 1974; Warth, 1986). The effects of wax and wax-based coatings on ethanol content, internal atmosphere and weight loss in apples and citrus fruit, and on the shelf-life of tropical fruits, have been reviewed in detail (Ben-Yehoshua, 1966; Blake, 1966; Bose & Basu, 1954; Cohen *et al.*, 1990; Dalal *et al.*, 1970; Eaks & Ludi, 1960; Hitz & Haut, 1938; Lawson, 1960; Magness & Diehl, 1924; Mathur & Srivastava, 1955; Nisperos-Carriedo *et al.*, 1990; Trout *et al.*, 1953). Wax coatings of vegetables, such as carrots, cucumbers,

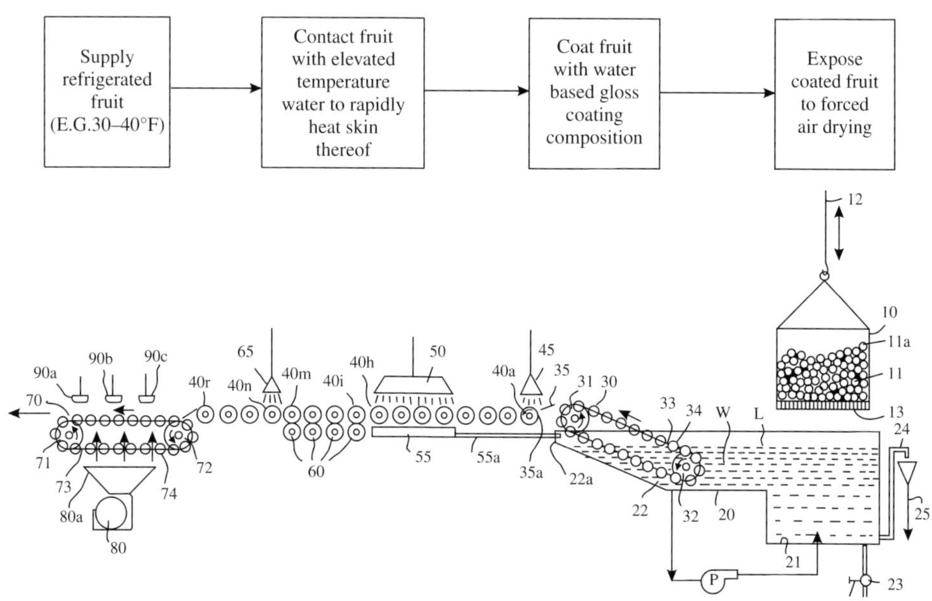

Fig. 2.1 Apparatus for wax, water-soluble or dispersible resin or oil glossy coating of apples and other fruits while the fruit remains refrigerated. (From Glasgow & Kraght, 1970.)

eggplants and pumpkins, have also been reported (Nelson, 1942; Wu & Salunkhe, 1972).

Other coatings which have been used in the past for fruits and vegetables based on hydrocolloids include: LMP (low-methoxypectin) to coat nuts and dried dates (Swenson *et al.*, 1953), powdered hydroxypropyl starch coating for prunes (Jokay *et al.*, 1967), amylose starch with required plasticizer for dates and raisins (Moore & Robinson, 1968), amylose ester of fatty acid and a layer of soy or zein protein for freeze-dried peas, carrots and apple slices (Cole, 1965), carboxymethyl cellulose (CMC) dust and starch for freshly cut pieces of fruits and vegetables (Mason, 1969) and chitosan and lauric acid (Pennisi, 1992) for apple slices.

Many of the coatings used today for fruits and vegetables are similar to those used in the past. As stated, wax coatings or those commercial coatings based on wax and/or polyethylene wax are listed elsewhere (Hardenburg, 1967; Kaplan, 1986; Krochta *et al.*, 1994). Methylcellulose (MC) has been used to coat fruits and prevent moisture loss. Hydroxypropyl cellulose (HPC) is unique in that it is a true thermoplastic that can be extruded into films from the molten state. HPC films are marketed by Watson Foods (West Haven, CT). These films are used to form pouches that allow processors to add pre-measured amounts of additives, such as colorants and vitamin pre-mixes, directly, without further handling.

Pears and bananas can be coated with CMC and fatty acid ester emulsifiers (Banks, 1984). This commercial coating was first called Tal Pro-long and later, Pro-long. It efficiently increases resistance to fungal rot in apples, pears and plums but is less effective at decreasing respiration rate and water loss in tomato and sweet pepper (Lowings & Cutts, 1982; Nisperos-Carriedo & Baldwin, 1988). Tal Pro-long has also been found to contribute to better flavor, and lower ethanol levels in Valencia oranges (Nisperos-Carriedo *et al.*, 1990).

A coating of similar composition, Semperfresh, contained a higher proportion of short-chain unsaturated fatty acid esters and was found to retard color development and retain acids and firmness in apples, and be responsible for storage life of citrus; however, it was not effective in retarding water loss in melons. Addition of waxes to Semperfresh improved the coating's shine (Curtis, 1988; Smith & Stow, 1984).

Carrageenan coatings were developed by Mitsubishi International Corp. (NY) for fresh produce (IFT, 1991). Other carrageenan coatings have been used to retard moisture loss from coated foods (Torres *et al.*, 1985). Sodium alginate- and gellan-based coatings were used to coat mushrooms for shelf-life extension and to prevent changes in texture during short-term storage (Nussinovitch, 1994; Nussinovitch & Kampf, 1992, 1993). More detailed research analyzed the relationships between hydrocolloid coatings and mushroom structure (Hershko & Nussinovitch, 1996, 1998b). Alginate and alginate–ergosterol, with or without emulsifier, were used to coat edible *Agaricus bisporus* mushrooms. The structure of the mushroom tissue was studied in detail since compatibility between the coating and the outer surface is important for coating success. The mushroom's tissue is porous and various ways of penetrating it, including minerals used as cross-linking agents and water-soluble polymers used for coating preparation, were examined. Surface tension of the coating solution is one of the important parameters to be considered since a reduction in its value can lead to better wettability of the surface. The edible coating contributed to better color, appearance and weight maintenance of the mushrooms with respect to their uncoated counterparts. A combination of alginate, ergosterol and Tween has been observed to be the best choice for maintaining size and shape of the coated mushroom (Hershko & Nussinovitch, 1998b).

Coating is designed to affect permeability to oxygen and carbon dioxide: thus, the coated fruit or vegetable becomes an individual package with a MA (Nussinovitch, 1997). Respiration of the commodity causes a depletion of oxygen and buildup of carbon dioxide. Care must be taken in designing the coating: if oxygen levels drop too low, anaerobic reactions will proceed, resulting in off-flavors and abnormal ripening (Kader, 1986). Ethanol and acetaldehyde concentrations in the tissue can be used as monitors for final and near-to final products of anaerobic respiration. Levels of oxygen below 8% decrease ethylene production, and carbon dioxide levels above 5% delay or prevent many responses to ethylene in the fruit tissue, including ripening (Kader, 1986). Therefore, if the coating can create these moderate MA conditions, a climacteric fruit will exhibit decreased respiration, lower ethylene production, slower ripening and extended shelf-life (Nussinovitch, 1997).

2.3.2 Meat, seafood and fish coatings

Meat is one of the main sources of protein in the Western diet. The animal meats that are most commonly eaten worldwide are beef, veal, lamb, pork, fowl, and less often, game animals. The proportion of protein, carbohydrate and fat in animals is extremely variable, depending on the type of animal and the particular cut of meat. In food preparation, most attention is given to the proportions of lean meat, fat, connective tissue and bone. These components directly influence the edibility of the meat and, therefore, the appropriate preparation method (Gates, 1981). Because meat is an attractive medium for bacteria, the holding conditions during aging must be carefully controlled to prevent the growth of putrefactive bacteria. Ideal aging conditions include a temperature of 1–2°C, a carbon dioxide atmosphere and 70% humidity for 3–6 weeks. Obviously, such storage is expensive. Meat coatings are designed to treat fresh or processed meats (Gates, 1981). A number of edible polysaccharide coatings based on alginates, carrageenans, pectin and starch derivatives have been reported for use with meats (Fig. 2.2). In general, hydrocolloid meat coatings create a barrier against moisture transport, reduce off-odor and drip, and retard the development of oxidative off-flavors.

Fresh meat was coated with a sodium alginate oligosaccharide solution cross-linked with calcium salt. Solutions were successively applied to the meat by either spraying or dipping. On this basis, the Flavor-Tex coating was developed. This formulation, which was sold commercially, included maltodextrin along with sodium alginate in the first solution, and CMC along with calcium chloride in the second (Earle & McKee, 1976; McCormick, 1975). Alginate Flavor-Tex coatings applied to lamb carcasses and beef cuts stored at 4 and 5°C, respectively, reduced moisture loss without significantly affecting total aerobic microbial counts on coated meat surfaces (Williams *et al.*, 1978).

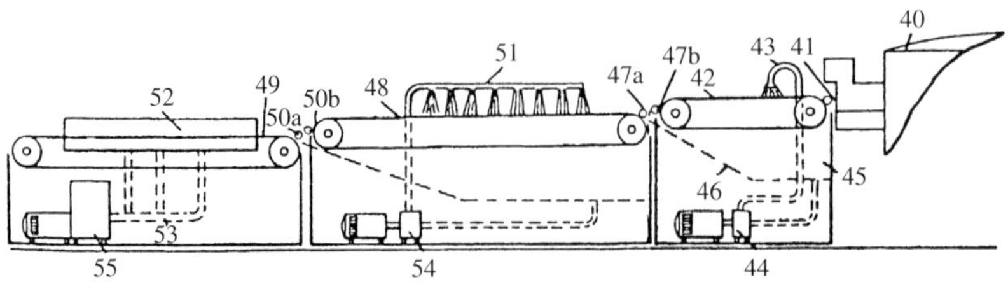

Fig. 2.2 Apparatus to apply alginate coatings to meat and fish, including one or more spray nozzles provided with means for forcing liquid through; tanks containing liquid are positioned relative to the conveyor system such that products are conveyed through the liquids in the tank. Drying is performed with blasts of cold or hot air. (From Hilgeland, 1964.)

In Japan, a polysaccharide edible film (Soafil) has found wide use in the meat industry as a casing for processed, smoked meats, such as ham and poultry products. The meat can be wrapped in edible film before soaking and steaming. The smoke flavor permeates the film during the smoke cycle, and the film dissolves during steaming. The finished ham has a smooth, lustrous surface and a good-textured bite on the surface, instead of the usual thick, rind-like texture. Yields are better because less moisture is lost during processing (Nussinovitch, 2000).

Carrageenan coatings have also been used to reduce off-odor development in chicken carcasses (Pearce & Lavers, 1949). Carrageenan, with or without soluble antibiotics, has also been tried on poultry meat and fresh chicken parts, especially with antibiotics found to be effective spoilage retardation agents; however, antibiotics in general are not approved for use on poultry meat as preservatives (Meyer *et al.*, 1959). Similarly, agar coatings with water-soluble antibiotics were found to be effective in extending the shelf-life of poultry parts stored at 2°C (Meyer *et al.*, 1959). Agar coatings were also used as a vehicle for the addition of $500\,\mu g\,ml^{-1}$ nisin (bacteriocin) to fresh poultry stored at 4°C in order to reduce contamination by *Salmonella typhimurium* (Natrajan & Sheldon, 1995).

More than 240 different species of seafood are consumed in the United States. Seafood may be divided into two general groups, fish and shellfish. Fish are aquatic animals with fins, found in fresh or salty water. Shellfish include crustaceans and mollusks. Coatings for seafoods are used to minimize freezer burn and weight loss, reduce off-flavor and maintain freshness and quality for longer periods. Flavor-Tex coatings (see earlier) have been found to be advantageous during frozen storage of red snapper and silver salmon (Ijichi, 1978). Carrageenan coatings have been used for shelf-life extension of fatty fish such as mackerel fillets (Stoloff *et al.*, 1948).

Seafood is among the most perishable of all foods. Fresh, raw or cooked seafood, depending on its freshness and variety, can only be stored in the refrigerator for a few days. Most seafood freezes well when packaged in moisture- and vapor-proof packaging, if it is frozen rapidly and stored at −18°C or lower. Seafood is cooked to improve its sanitary quality and to develop flavor, texture and tenderness; because it has little or no connective tissue, very little cooking is required to produce tenderness. If seafood is cooked at too high a temperature or for too long, it dries out and loses its tenderness and flavor. Raw seafood has a watery, translucent coloring and turns opaque when cooked. In general, seafood is sensitive and thus needs to be coated to extend its shelf-life. Dextran coatings, for example, have been applied in the form of aqueous solutions or dispersions, to peeled and unpeeled shrimp (Toulmin, 1956a, b), fish and meat products, such as ham, sausage and bacon (Toulmin, 1957) to improve their freshness, flavor and color when refrigerated or frozen.

Seafood can be protected from dehydration and oxidation during freezing by first inactivating microorganisms and enzymes by heating in water, followed by immersion in an aqueous dispersion of sodium alginate and native starch, oxidized starch or dextrins, and sometimes even vegetable oil. The coating is gelled by introducing it into a calcium chloride solution. Shrimp, mackerel and kingfish treated in such a manner retained their flavor, texture and color after storage for 3 months under freezing, in comparison to non-coated products which did not (Earle & Snyder, 1966).

MC and hydroxypropylmethyl cellulose (HPMC) have been used in edible coatings for seafood entreés and meat parts, and were found to provide a moisture barrier during cold storage and thermo-gelation (MC), and increase cooked product yields (HPMC) (Anon, 1993; Bauer & Neuser, 1969). Developments in edible films include: packaging of beef cubes in a collagen wrapping; milk-protein-based films; grain-based films which can be used as animal feed; and HPMC use in films (Rice, 1994).

2.3.3 Coatings for fried products and oil resisters

Frying is a unit operation used to alter the eating quality of a food. A secondary consideration is the preservative effect that results from thermal destruction of micro-organisms and enzymes, and the reduction in water activity at the surface of the food (Fellows, 1990). Fats and oils are cooking mediums. Frying cooks foods quickly because of the high temperatures involved. Moreover, many people enjoy the flavor and texture of fried foods. However, these foods are higher in calories than foods cooked with water or by other methods. Thus, fried foods with lower calorie contents are in demand and can be achieved, at least partially, by the use of appropriate coatings (Gates, 1981).

An edible coating system, Fry-Shield™, developed and patented by Kerry Ingredients (Beloit, WI) and Hercules (Wilmington, DE), can be used to create a film around a food product, by treating calcium with LMP. The resultant film reduces the amount of fat uptake during frying by 20–40% (Anon, 1997).

Gellan gum films can also act as barriers to oil absorption in battered and fried foods. Gellan gum-based coatings have been used for several years in Japan and other Asian countries with tempura-type fried foods. MC and HPC manufactured by the Dow Chemical Co. (Midland, MI) have been used to decrease oil absorption during the frying of french fries and onion rings (Anon, 1997).

Of all the films available to the flexible packaging industry, the edible and water-soluble films are the most fascinating. Their uses are nearly unlimited. Specific examples include films produced from water-soluble cellulose derivatives. The material is cast from an aqueous suspension and is soluble within the 0–55°C range (Anon, 1977). These films are excellent resisters of oil and grease and have favorable mechanical properties. Colors and flavorings can be incorporated directly into the film.

Delmar Chemical Company introduced the first commercial packaging using a film based on high-amylose cornstarch. The American Maize Products Company

(Hammond, IN) produced *Ediflex*™, a hydropropylated amylomaize starchy material with excellent resistance to oil and grease. In addition, its film is an excellent barrier to oxygen, carbon dioxide and nitrogen. *Ediflex*™ can be successfully printed and heat-sealed, and machined on most high-speed packaging equipment. For lamination, consideration must be given to the use of non-aqueous adhesives due to the sensitivity of the film. *Ediflex*™ can protect meat products during frozen storage and subsequently be dissolved during thawing and cooking (Gennadios *et al.*, 1997).

2.3.4 Nut and peanut coatings

Fresh, roasted or candied nuts are popular snacks. Many types of nuts exist, including almonds, Brazil nuts, cashews, coconuts and hazelnuts, among others. Their flavor and texture contribute to many dishes. In general, nuts (except chestnuts) differ in their flavor, but can be used interchangeably. In many cooking recipes, nuts are added right before serving to retain their crunchiness (Gates, 1981).

For coating purposes, it is recommended that the size of the nuts be uniform. Therefore special sorting processes should be performed. Specifically, 38–42 nuts per ounce are shelled and sorted before hot-air roasting with their skins intact and a period of *naturalization* for 12–24 h. Brittle skins are removed by passing the nuts through rollers: the skins are removed by suction and pelletized for animal feed. After reaching the required uniform size, the nuts can be coated in a coating drum with gum, starch and a spice mixture, followed by a second roasting, again without the addition of any fat or oil. Packaging is in laminated foil packs which are flushed with N_2 to maximize shelf-life (Byrne, 1982).

LMP plasticized with glycerin and gelled with calcium chloride has been reported as a coating for almonds to bind salt uniformly to the kernels without applying an oily surface (Swenson *et al.*, 1953). A coating which included antioxidant was found to slow the rate of rancidity in coated nuts (Andres, 1984). Coatings based on wheat gluten, dextrin, modified starches and gum arabic (Fig. 2.3) have also been invented for improving the adherence of salt to and flavoring in nuts (Arnold, 1963; Daniels, 1973).

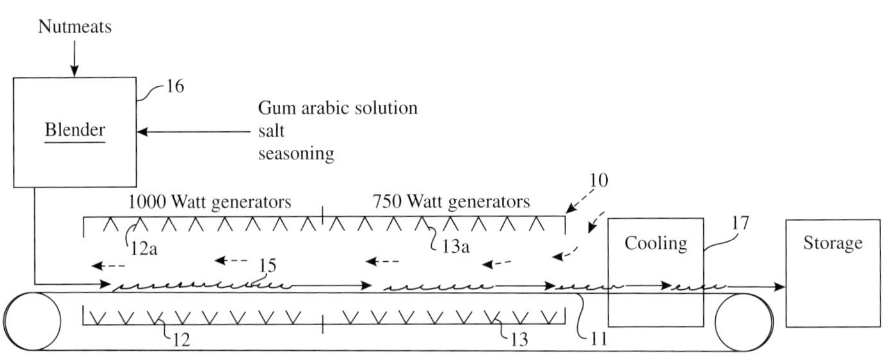

Fig. 2.3 Gum arabic coating of nutmeats. The process involves roasting and coating nutmeats without breaking; it includes coating deliberately flavored, fragile, whole pecan nut halves, in their raw shelled state, with a protective coating of gum arabic, salt and spices, and passing the coated nuts through a controlled infrared tunnel or oven at a regulated speed. (From Arnold, 1963.)

A coating from a hydroxypropyl derivative of high-amylose starch was effective in delaying oxidative rancidity in almonds (Jokay *et al.*, 1967).

Peanut, also called *groundnut* because it develops underground, is a pea and belongs to the legume family. Peanuts have higher protein content and quality than any other nut, and they are often the least expensive as well. The skin and germ of the peanut are good sources of thiamin. However, despite peanuts' slightly higher nutrient content, commercially packed peanuts often come with the germ removed, because it is thought to contribute some bitterness to the taste. Peanuts are available in many forms (Gates, 1981), of which roasted is one of the most popular. The shelf-life and quality of roasted peanuts is determined by the extent of lipid oxidation, which can be controlled with a suitable coating. A specific example is coating peanuts with HPC and zein at 4, 7, and 9% of their initial raw material weight and incubation at 30°C. Such a coating can delay hexanal production. Thus enhancement of shelf-life is observed but it is not dependent on the coating level (Ramon *et al.*, 1996).

2.3.5 Confectionery coating

Confections are sugar products produced by boiling a solution of sugar, flavoring and coloring to the required concentration and then setting the mix in moulds without crystallization (Sutherland *et al.*, 1986). Confectionery items may also include chocolate pieces, buttercreams, sugar-coated, roasted peanuts and hard gums (Alikonis, 1979). Fillings such as sherbet powder or semi-solids may be incorporated. Quality is largely determined by personal preference; for example, sweets produced for children are likely to be of a garish color that is unlikely to appeal to adults (Balke, 1977). High-quality sweets of this type are also likely to be more clearly molded and to have fewer broken pieces (Sutherland *et al.*, 1986). Extrusion cooking is one of the processes used to produce a gelatinized chewy confection product (for example fruit gums and licorice) from a mixture of sugar, glucose and starch. The heat gelatinizes the starch, dissolves the sugar and vaporizes excess water, which is vented from the machine under vacuum. Colorings and flavorings are added to the elasticized material and, after mixing, it is cooled and extruded (Fellows, 1990). The confection's texture is adjusted from soft to elastic by controlling the formulation and processing conditions (Fellows, 1990). The shape is changed by changing the die, and a variety of flavors and colors may be added. These different combinations permit a very large variety of potential confection products, all using the same equipment. Product uniformity is high, no final drying is required, and start-up and shutdown are rapid (Fellows, 1990).

Confectionery products are coated to extend their shelf-life: the coating serves to retard oxidative rancidity, staling, oil leakage and moisture absorption. In addition, the coating can contribute to the gloss and aesthetic appeal of the confection. Confection gum coatings or films positioned between confection layers to inhibit lipid migration can be based on starches, dextrin, MC, gum arabic, or other gums or their combinations. A few examples are briefly mentioned in the following paragraphs.

Advantages of low-viscosity starches for confectionery coating have been described, with particular reference to the Crystal Gum range of tapioca dextrins from National Starch & Chemical (Germany). These specialty starches offer an alternative to gum arabic and are suitable for the manufacture of chocolate-coated products (e.g. raisins) and for the glazing of sugar coatings (Bull, 1996).

Use of tapioca dextrins (crystal gum products) as a replacement for gum arabic in dragee-type confectionery is described. Covered aspects include: comparison of cost and functional properties of tapioca dextrins and gum arabic, film-formation properties, use of tapioca dextrins in hard sugar coatings for nuts and use in chocolate coatings (Anon, 1995). Chocolate or compound coatings (in which cocoa solids and hardened vegetable oils replace cocoa butter) are used to enrobe confectionery, ice cream and baked goods. The principal ingredients in a coating are fat and sugar. Corn syrup, flavorings, fat-soluble colorings and emulsifiers are also added to achieve the desired properties (Fellows, 1990). Using semi-solid edible coatings that contain high-methoxy pectin (HMP), acacia gum, high-fructose corn syrup, dextrose, fructose and sucrose can inhibit lipid migration in chocolate-enrobed products. Such coatings, containing gums and sweeteners, were examined for their lipid-barrier abilities, sensory properties, adhesion to chocolate, viscosity and water activity. At a water activity of 0.5 and a thickness of 0.5 mm, no migration of oil was detected after 47 days at $30 \pm 1°C$. Uncoated products suffered from an equivalent migration of $11.5\,mg\,cm^{-1}$ trilinolein after 1 day at $21 \pm 2°C$ (Brake & Fennema, 1993).

In the case of coatings (0.5 mm thickness) positioned between chocolate and peanut butter, no significant differences between coated and non-coated food samples were observed except for the color of the coating (Brake & Fennema, 1993). Retardation of lipid migration in the case of crushed nuts within a chocolate coating is the topic of another manuscript (Nelson & Fennema, 1991). For a model system, MC films were chosen and exposed to peanut oil at 30°C and a 5 psig pressure differential. If imperfections were avoided, less than 1.0 μg linoleic acid equivalents passed through the coating. The thicker the film, the easier its conception by a sensory panel; the thinner the film, the further it was from perfect (Nelson & Fennema, 1991).

Koenigsberger marzipan products are small and round, oval or rectangular flan-base-shaped shells made from marzipan, filled with fondant, jam or fruit preparations, and flamed in a special flaming oven. To minimize its stickiness to the inside of the packaging material, the flamed marzipan is coated with a solution of gum arabic (Anon, 1980).

2.3.6 Glazes

Many natural gums are used in the baking industry for diverse purposes ranging from modifying dough properties to stabilizing fillings, icings, meringues and glazes (Klose & Glicksman, 1972). The appearance of glazes on baked goods contributes to their acceptability and appeal. Glazes on sweet bakery products are basically sugar and water with other ingredients, such as whipping agents, flavor, shortening, milk solids and stabilizers (Klose & Glicksman, 1972). A common problem with glazes is loss of water, resulting in sugar crystallization and cracking, chipping, or sweating of the glaze. This problem can be minimized by using a low-moisture hot glaze containing hydrocolloid at a concentration of 0.5–1.0% of the sugar level. The gum functions as a glaze thickener, increases its adherence to the baked goods, imparts quick setting properties, provides flexibility to prevent chipping, and reduces glaze melting (Klose & Glicksman, 1972). The gums used in the glazing industry are agar, due to its ability to withstand high temperatures; blends of corn syrup and an equal weight of 2% carrageenan (brushed warm to give a highly glossy glaze); and gum arabic and CMC (to retard sugar crystallization). Edible films can also act as glazes to enhance the appearance of baked goods. A good example is a wheat-gluten coating that replaces the traditional egg-based coating (Anon, 1997). Opta Food Ingredients (Cambridge, MA) market this particular edible film under the name OptaGlaze™. In addition to supporting glossy appeal, this film eliminates possible microbial problems associated with raw egg-based coatings and provides some barrier properties against moisture loss (Anon, 1997).

2.3.7 Lightly processed agricultural products

Lightly processed agricultural products present a special problem to the food industry and to scientists involved in post-harvest and food technology research (Baldwin *et al.*, 1995). Light or minimal processing includes cutting, slicing, coring, peeling, trimming, or sectioning of agricultural produce. These products have an active metabolism that can result in deteriorative changes, such as increased respiration and ethylene production. If not controlled, these changes can lead to rapid senescence and general deterioration of the product. In addition, the surface water activity of cut fruits and vegetables is generally quite high, inviting microbial attack, which further reduces product stability. Methods for controlling these changes are numerous and include the use of edible coatings. Also mentioned in this review are the coating of nut products, and dried, dehydrated and freeze-dried fruits. Technically, these are not considered to be minimally processed foods, but many of the problems and benefits of coating these products are similar to coating lightly processed ones. Generally, the potential benefits of edible coatings for processed or lightly processed produce are product stabilization and consequent shelf-life extension. More specifically, coatings have the potential to reduce moisture loss, restrict oxygen entrance, lower respiration, retard ethylene production, seal in volatile flavors and carry additives that retard discoloration and microbial growth (Baldwin *et al.*, 1995).

2.3.8 Rice coating

Rice is the most important crop: almost half of the world's population depends on rice as the principal food in their diet. White rice or refined rice is rice from which the bran and germ have been removed; brown rice contains the bran and germ. White rice is often polished with glucose and talc to improve its luster, hence its name, polished rice. Most people prefer the color and bland flavor of white rice. A technique to delay the release of soluble nutrients from rice during cooking (in excess water) until it reaches the digestive tract has been described. It involves coating the rice (2 mm coating thickness) with a polymer system based on starch or cellulose that is insoluble at 100°C but soluble at 37°C. Tests on thiamin-enriched Patna rice showed that the best coating comprised 1.2% (by wt.) MC, 3.6% HPMC, 28.5% ethanol (95%) and

66.7% water. Nutrient level varied insignificantly in rice stored for less than 8 months. The aim of the process was to produce a blend of non-fortified rice, fortified rice (at 1%), and a rice with the recommended nutrient levels. Other possible applications are fortification of other whole grain cereals or other foods (Hannigan, 1983).

Another idea is to place a layer of edible film over rice, to cover the film with sauce and then to freeze the product. The film is dissolved upon heating. Such a film can help maintain the crispiness of fried foods and may also extend the shelf-life of fresh fruits and vegetables by eliminating moisture accumulation. However, the moisture absorbed by the film may cause its dissolution (Mitsubishi International Corp., Fine Chemicals Dept., NY). Other edible films were prepared using a combination of rice protein concentrate and the polysaccharide pullulan. The protein–pullulan mixture, with up to 50% protein concentrate, can be cast on a glass plate into films with a tensile strength of ~18 MPa and water vapor permeability (WVP) of 40 g mil m^{-2} day^{-1} $mmHg^{-1}$. Film strength and WVP were improved by the addition of small amounts of propylene glycol alginate (PGA) under alkaline conditions. Oils were also incorporated into the film for improved water vapor resistance (Shih, 1996).

2.4 Special biopolymer-based coatings

2.4.1 Polyvinyl alcohol (PVA) and polyethylene oxide (PEO) coatings

PVA is a white, powdered synthetic resin which is manufactured by the hydrolysis of polyvinyl acetate. It is the only polyhydroxyl polymer that is readily soluble in water. The PVA is used in adhesives, in the manufacture of automative safest glass, in textile sizes and finishes, in paper sizes and coatings and as a binder for ceramics, foundry cores, non-woven fabric and various pigments (Davidson, 1980). Unsupported film cast from water solutions of PVA and plasticizer is transparent, tough, tear-resistant and puncture-resistant. It provides a unique combination of water solubility, gas-barrier properties and resistance to oils, grease and solvents (Davidson, 1980). Specific water-soluble PVA film produced at the requested predetermined thickness of 3.8×10^{-5} m has been described (Anon, 1977). It is a translucent heat-sealable film which is soluble in both hot and cold water. Thus it provides a wide spectrum of options for food coatings. PVA films are poor barriers to water vapor but excellent barriers to oxygen and grease (Anon, 1977).

PEO is a very high-molecular-weight resin that was first investigated in the 1930s. In 1958, Union Carbide began the first commercial production of these resins. Since then, two Japanese companies have begun producing PEO. These resins are made commercially by the catalytic polymerization of ethylene oxide in the presence of any of several different metallic catalyst systems. They are available with average molecular weights of as low as 200 and as high as 5 million (Davidson, 1980). Above their melting point, these resins exhibit rheological properties typical of thermoplastics and can therefore be formed into various shapes using conventional thermoplastic processing techniques. Films are produced by calendering or blown-film extrusion techniques, usually in thicknesses of 38.1–76.2 μm, and have very good mechanical properties combined with complete water solubility (Davidson, 1980). PEO films have the best solubility characteristics of all water-soluble films. They can be oriented to provide good burst properties and will elongate up to 600% with no return. PEO films are used for soluble laundry bags and for packaging a variety of industrial additives. They are less suitable for foods than PVA and some cellulosics; however, they are used as a water-soluble film for packaging dyes (Anon, 1977).

2.4.2 Pullulan-based films

The fermentation of starch syrups and crude sugars by *Aureobasidium* (*Pullularia*) *pullulans* produces a viscous extra-cellular polysaccharide, pullulan, at a 70% yield. This biopolymer has potential uses in the food industry, for example, as a spray-on coating for fruits and vegetables, and for packaging films and containers. It can be used as an ingredient in low-calorie dietetic foods and drinks, imparts satisfactory texture, viscosity, dispersibility and moisture retention, and inhibits fungal growth. Pullulan is supplied as a fine white powder which is odorless and tasteless. It dissolves readily in cold water to give a colorless liquid. Pullulan is advocated for its flexible, water-soluble, biodegradable films with high impermeability to O_2, its suitability for use in foods and the plastic character of certain of its derivatives (Jeanes, 1976; Yuen, 1974).

Films based on blends of pullulan, sodium caseinate, and soy protein isolate (SPI) or a peptide (Hinute PM) were investigated by Adachi *et al.* (1995). The parameters tested were: film surface characteristics, solubility and diffusion coefficients of O_2, and the influence of SPI on oxygen permeability and critical surface tension. The structure of those films became rougher with the addition of SPI. Addition of the peptide had no such effects on the surface or permeability of pullulan-based films (Adachi *et al.*, 1995).

2.4.3 Chitosan-based films

Chitosan is the principal product of the alkaline hydrolysis of chitin, a main constituent of the exoskeleton of crustaceans such as crabs, lobsters and shrimps (Arai *et al.*, 1968). The gelling properties of chitosans enable a wide range of applications, the most attractive being gel entrapment of biochemicals, plant embryos and whole cells, micro-organisms, and algae (Pegg & Shahidi, 1999). Chitosan has also been suggested for use in food products and coatings, although for the time being it has not yet been approved for food use in the United States or Europe; it has, however, been approved in Canada (Baldwin, 1999). Chitosan films are semipermeable with the property of modifying the internal atmosphere of the coated vegetative produce; they can also serve as fungus- and pathogen-growth inhibitors (El Ghaouth *et al.*, 1991).

Chitosan films have also been used in membrane separation, preservation of fruits and vegetables, and meat casings. Relationships between the chain flexibility of chitosan molecules in solution and the physical properties (gel swelling index, maximum melting temperature, and tensile strength) of their casted films were studied in order to manipulate the conditions to tailor the physical properties of the films being produced. The chain flexibility of chitosan molecules in solution was manipulated by using chitosans with different degrees of de-acetylation (DD) in media with different pHs and ionic strengths, and different solvent systems. Results showed that the gel swelling index is independent of media pH, and decreases with increasing DD of the chitosans used, whereas maximum melting temperature and tensile strength increase with the increase in DD of the chitosan molecules. The differences were attributed to different crystalline regions in the films (Rong Huei Chen *et al.*, 1994). The ability of chitosan and cross-linked chitosan films to transmit water vapor and oxygen, and to resist grease, has been studied (Kittur *et al.*, 1998). In addition, typical physical properties, such as tear resistance and tensile and burst strength, were evaluated. Cross-linked chitosan films were less strong and more permeable. WVPs were influenced by water activity and resembled the behavior of other films based on hydrophilic polymers (Kittur *et al.*, 1998). Another study analyzed the physical properties of laminated films prepared from casting chitosan–lactic acid films on pectin films. Plasticizers such as lactic acid or glycerol can be used to achieve dynamic mechanical properties similar to those of pectin films alone. The laminated films had greater storage and loss moduli than the films made up of only chitosan. Lamination did not affect water vapor permeation of chitosan or pectin films (Hoagland & Parris, 1996).

Chitosan films plasticized with glycerin were studied for their coefficients of oxygen permeability (OP), WVP and ethylene permeability (EP), tensile strength and elongation at break (E). Storage for up to 12 weeks was considered. During the first two weeks of storage, an initial decrease in permeability was detected, followed by mean OP and EP remaining constant (which was not expected), while mean WVP decreased. Storage decreased film strength and changed its brittleness (as expected). The chitosan films were said to be resistant to O_2 permeation and have relatively low water vapor characteristics. Thus the addition of plasticizer is important for ease of film production but can have a negative effect on the film's barrier properties (Butler *et al.*, 1996).

Chitosan–MC films including preservatives (sodium benzoate or potassium sorbate) exhibited antimycotic activity against *Penicillium notatum* and *Rhodotorula rubra*. The incorporation of preservatives had no effect on the strength and elongation of the tested film. More than 40% of the preservatives were released from the film at 25°C to a glycerol–water mixture in 30 min. Maximum release can reach values of up to 65% at that temperature. Reduced temperatures decrease the rate and quantity of released preservative. It could be that the rate is dependent on ionic interactions between $-COO^-$ of the preservatives and $-NH_3^+$ of the chitosan (Chen *et al.*, 1996).

The clarity of the film is an often requested optical property. Chitosan–HCl and HMP with glycerol as an added plasticizer provides this property, along with good dynamic mechanical properties. The pectin stabilizes the chitosan-based film and prevents shrinkage and opacity. In such systems, starch can serve as a good stabilizer; however, it has a detrimental effect on the dynamic mechanical properties of the film (Hoagland, 1996).

2.4.4 Gellan gum-based films

Gellan gum is an extra-cellular polysaccharide produced by the bacterium *Sphingomonas elodea*, designated S-60 or PS-60 in the early literature. Gellan is a linear, anionic hetero-polysaccharide based on tetrasaccharide repeat units. The native polysaccharide is partially esterified with L-glycerate and O-acetate. The normal commercial product is a de-acetylated polysaccharide produced by alkali pre-treatment of the broth prior to alcohol precipitation (Hill *et al.*, 1998).

To study the structure of gellan gum, thin films in the presence of Na^+, K^+, Ca^{2+} or no salts were deposited on highly oriented pyrrolytic graphite (HOPG) by means of scanning tunneling microscopy (STM). The gellan was seen to have a double-helical structure which, in the presence of K^+, associates into cation-mediated aggregates. The arrangement was dependent on interactions between the gellan film and the pyrrolytic graphite, and on the solvent in the initial gel deposited on the substrate before drying. The length of the gellan strands depended on the type of cation, and corresponded to the number of links and strength of the gel when the zipper model was considered (Nakajima *et al.*, 1996).

Gellan coatings of garlic bulbs were studied. The coatings served as a barrier to moisture loss. Incorporation of ingredients that can be found naturally in garlic skin, or are chemically similar to these, into the gum solution before coating, improved adhesion of the film to the surface of the coated commodity (Fig. 2.4). Adhesion strengths were about three times higher than those recorded for a film made of gum and cross-linking agent alone. Distances between the film and the vegetable were measured using image-processing and they could sometimes be reduced by varying the film composition (Nussinovitch & Hershko, 1996; Nussinovitch *et al.*, 2000).

2.4.5 Starch inclusion into formulations and films

Starch is the major storage polysaccharide of higher plants where it occurs in organs such as seeds, tubers and roots, and also in smaller quantities in stems and leaves. It exists as water-insoluble, roughly spherical granules whose shape, size and size distribution are characteristic of the particular plant species. Typical major size dimensions are $\sim$2–100 μm. Starch granules dispersed in water exhibit a limited degree of swelling, with the process being exothermic. Irreversible swelling of granules occurs when the dispersion is heated above the gelatinization temperature. Gelatinization results in a loss of molecular orientation and a breakdown of the crystalline structure. Swelling of the granules leads to solubilization of the amylose. Heating results in porous amylopectin-based granules suspended in a hot amylose solution. On cooling or at sufficiently high polysaccharide concentrations, the sample forms a turbid viscoelastic paste (Hill *et al.*, 1998).

Starch can be included in coating solutions and in encapsulation preparations in different proportions to change their basic properties. Oxidized starches originating from corn and amaranth (*Amaranthus paniculatus*) are used as substitutes for gum arabic in encapsulation. Information on their encapsulation properties can be found elsewhere (Chattopadhyay *et al.*, 1997).

Starch films can be used for packaging especially when biodegradability is important (Zoebel, 1991). Extrusion has been observed as one procedure to study the physical properties of starch films. The deformation of thin starch films is very important. Films were cast from solutions of maize (corn) grits extruded at different temperatures. The initiation of deformation zones in the granular remnants was

(a)

(b)

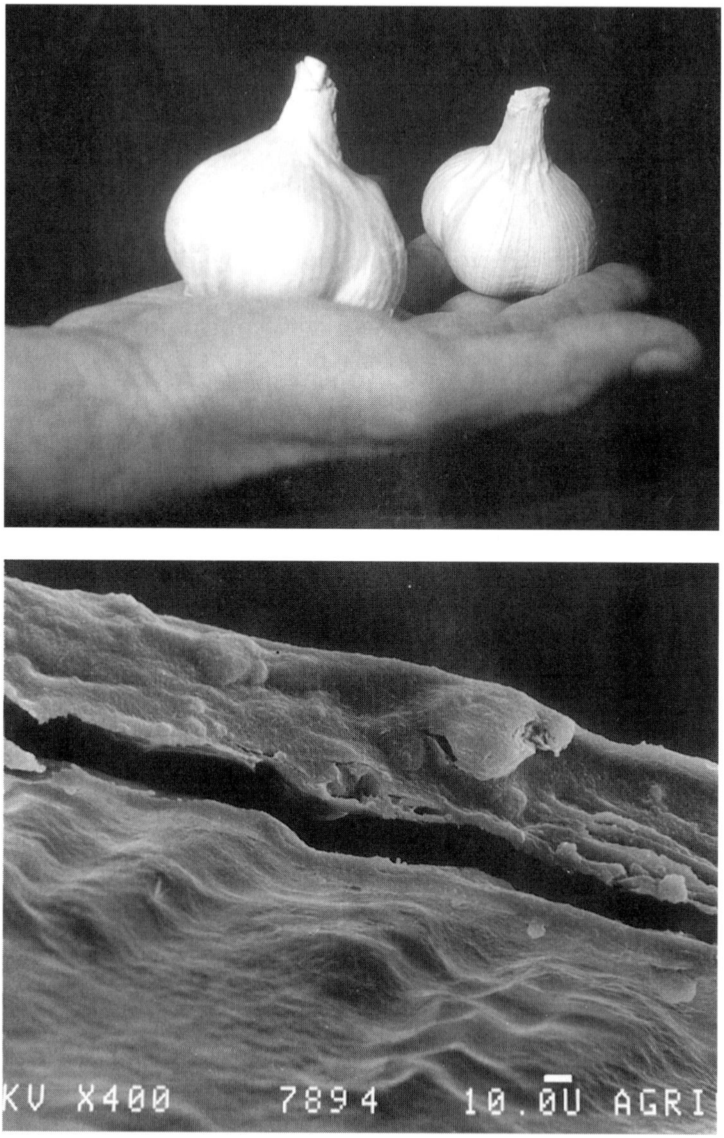

Fig. 2.4 (a) Gellan-coated garlic bulbs. (b) Gellan–sitosterol-coated garlic. The distance between the film and garlic is equal to or smaller than 20 μm. (From Nussinovitch & Hershko, 1996, with permission from Elsevier Science.)

found in films made from both unextruded and extruded grits at low temperature. Extrusion at high temperatures resulted in more homogeneous films with higher strains needed to form deformation zones. Maximum strength was observed at an extrusion temperature of 140°C. At higher temperatures, degradation may have been involved in lowering the strain needed to form degradation zones (Warburton *et al.*, 1993).

Blends of starch and MC or micro-crystalline cellulose, with or without polyols, were extruded, hot-pressed and conditioned at different humidities. The physical properties of the films were investigated. An increase in polyol content or water produced a considerable increase in %E and a decrease in strength. High cellulose content increased the strength and decreased the water vapor transmission. Crystal development decreased water and gas permeabilities (Psomiadou *et al.*, 1996). Other blends of starch and gelatin also present interesting future possibilities for the food industry (Arvanitoyannis *et al.*, 1997).

The achievement of well-layered starch products depends, among other things, on their ability to include other ingredients or to interact with them. Thus, starch–oil relationships are important. A starch–oil composite (film) contained 1–10-μ oil droplets that were stably entrapped within (and see Chapter 3). Their size and distribution was determined by the oil-to-starch ratio, and the number of times the formulation was passed through the steam cooker and steam pressure (Eskins *et al.*, 1996).

Another starch-based film that achieved a commercial success is oblate – a Japanese-originated invention. It is a Japanese film based on starch. It is made from rice starch to which a small amount of vegetable gum is added (Anon, 1977). The dilute paste is subsequently drum-dried at 102°C on heated rolls. It creates thin films and needs to be stored at the appropriate controlled humidity to avoid brittleness.

2.4.6 Mesquite-based films

The mesquite tree is found in arid environments. To most Americans, mesquite usually just means charcoal. However, the pods have a unique flavor, and gum can be produced from the seeds. Edible mesquite films can be produced from the mesquite gum. The dependency of their creep compliance, when adsorbed at a liquid paraffin–water interface, on sodium or calcium chloride concentration, pH and gum concentration was investigated. The films were found to be viscoelastic if a short aging time was used, as against longer aging times, which yield only elastic nature. The lowest values of the rheological parameters were recorded at pH 7.0. Higher values were achieved when sodium or calcium chloride was added to the aqueous phase, with sodium contributing to a more intense effect (Vernon Carter & Sherman, 1981).

2.4.7 Edible films based on cellulose and their properties

Cellulose derivatives can be prepared by treating alkali cellulose with either methyl chloride to form MC, propylene oxide to form HPC (a side reaction in which propylene oxide reacts with water to form a mixture of propylene glycols also occurs, but can be minimized by keeping the water input as low as possible), or sodium chloroacetate to form sodium CMC. In this last case, a side reaction, the formation of sodium glycolate, also occurs (Stelzer & Klug, 1980). Mixed derivatives such as HPMC can be formed by combining two or more of these reagents. Of the many possible derivatives investigated and manufactured, CMC, MC, HPMC and HPC are utilized in the food industry in addition to modified forms of cellulose, which have been found to have useful functional hydrocolloidal properties and significance in several food applications. CMC is the most important cellulose-derived hydrocolloid. It is an anionic polymer which is important for thickening applications and is able to react with charged molecules within specific pH ranges (Ganz, 1977; Hercules Inc., 1978; Stelzer & Klug, 1980).

HPC, a non-ionic cellulose ether, is soluble in water below 40°C and in polar organic solvents (Butler & Klug, 1980). CMC is an acid-hydrolyzed, pure α-cellulose material which has thickening and water-absorptive properties and is used in frozen and dairy food products. MC and HPMC, which have emulsifying properties, form flexible films and contribute to freeze/thaw stability.

Polyethylene glycol 400 (PEG 400) has been used to change mechanical and other properties of MC. PEG is generally used as a coating, a binding and plasticizing agent, and/or a lubricant in food tablets. It can also be used as an adjuvant and bodying agent in non-nutritive sweeteners listed as GRAS; and as an adjuvant to disperse vitamins and/or mineral preparations. Its oral LD_{50} in rats is 30–50 g kg^{-1} for molecular weights of 200–9500 (Powell, 1980). As an additive to coating formulations, PEG has the potential ability to increase strength and flexibility, as well as increase coating permeability to water vapor and gases. Several edible films based on MC and PEG 400 were prepared and stored until equilibration at different relative humidities. Tensile stress–strain curves showed very different behaviors as a function of PEG 400 and relative humidity. Tensile strength strongly depended on relative humidity and then on water content, more than on PEG 400 content. In contrast, elongation was dependent on both water and PEG 400. These differences correspond to the glass transition of the polymer, which affects the elongation more than the tensile strength. However, from differential scanning calorimetry (DSC) measurements, it appears that PEG 400 has little or no compatibility with the MC matrix (Baldwin, 1999).

MC and PEG 400 were used to prepare edible films, the moisture sorption isotherms of which were determined at 10 and 25°C. The equilibrium moisture content of these films was very low at low water activity and increased sharply above a water activity of ~0.8. Increasing PEG content at higher water activity resulted in increased moisture content. Moisture content or WVP was not directly related to the temperature effect (Ayranci, 1996).

Another report (Greener & Fennema, 1993) deals with the effects of different plasticizers at 30% (dry basis) on the physical properties of MC films. In general, they increased the spacing of the crystal lattice (except for propylene glycol), and significantly increased OP and WVP as compared to their counterparts. PEG 400 had the greatest effect on OP, while glycerol had the greatest effect on WVP. Tensile strength of MC films decreased as a result of plasticizer addition (except in the case of PEG). Glycerol and PEG were the most effective plasticizers for MC (Greener & Fennema, 1993). *Klucel*, developed by Hercules Chemical Co., is a company tradename for HPC. The film is produced by a patented extrusion process. These films have been used experimentally for injection-molding of pharmaceutical capsules, and blow-molding of bottles, and have been tested for their capabilities of running on automatic packaging machinery (Anon, 1977). Hydroxyethyl cellulose (HEC) can be cast in a manner similar to cellophane, and yields a film with properties analogous to the regenerated cellulose (Baldwin, 1999).

Foods can be protected using edible or non-edible films to limit vapor or water transfer and to allow selective gas transfer. Films that contain hydrophobic substances can limit water vapor transfer, but film-forming agents such as proteins or cellulose derivatives must be used to improve the mechanical properties of the film.

The moisture barrier properties of two hydrophobic materials, paraffin oil and paraffin wax, were studied as a function of the characteristics of the technique employed to prepare hydrophobic films with a constant paraffin content. Three cellulose derivatives with different polarities and porosities were used as supports to prepare films by several techniques. Results showed that the ability of the hydrophobic substances to retard moisture transfer depends on the homogeneity of their final repartition in the matrix and/or on the surface. Highly heterogeneous systems (emulsion and dipped filter paper) represented the least efficient films in terms of retarding water movement, independent of the nature of the substance, as seen by scanning electron microscopy (SEM). The most efficient system, in terms of WVP, was obtained using cellophane films with homogeneous repartition of paraffin wax, independent of thickness and relative humidity. Cellophane films prepared with paraffin oil were more permeable than those prepared with wax (Martin Polo *et al.*, 1992).

2.4.8 Cellulose and proteins

A combination of celluloses (MC and HPC) and proteins (corn zein and wheat gluten) for the production of edible films could be useful when their permeability to gases needs to suit the early requirements of some products. WVPs of edible films were higher than those of plastic films, whereas their CO_2 and O_2 permeabilities were lower than those observed for plastic films. Permeability can be influenced (increased) by combining plasticizer within the formulation, while lipid addition to HPC films has the opposite effect. Gas permeabilities relied linearly on film thickness (Hyun & Chinnan, 1995). Other combined edible films based on MC, PEG and paraffin wax were studied for their tensile strength and water-barrier efficiency. In addition, the effect of different surfactants and drying conditions on a limited range of physical properties was detected. Among glycerol monostearate, acetate glycerol monostearate, citrate glycerol monostearate, sorbitol monostearate, glycerol monostearate (pure) and polyoxyethylene sorbitan monooleate, pure glycerol monostearate contributed mostly to the production of films with the lowest water vapor transmission rates and the greatest resistance to mechanical impact. Drying is an effective procedure for achieving better mechanical and barrier properties (Debeaufort & Voilley, 1995).

2.5 Special uses and aspects of hydrocolloid films

2.5.1 Biodegradable plastics

Edible films and coatings provide a useful alternative to plastics intended for direct contact with foods. Coatings may increase shelf-life, by retarding the migration of moisture, gases, oils and fats. They can also improve structural integrity or handling properties. Antimicrobials can be added to edible films to retard growth of yeasts, fungi and bacteria; antioxidants or ingredients that prevent color changes can also be added. Films can be coated onto foods, but they can only exist as continuous layers between compartments of the same food product. Possible applications for edible films and coatings include frozen foods and bakery, dairy and meat products (Myllarinen *et al.*, 1997).

Biodegradable plastics can be based on corn or cornstarch, potato, wheat or rice starch, chitosan, plastics derived from lactic acid fermentation of corn feedstock, carrageenan, polysaccharides and pullulan polymers. Biodegradable plastics can be used as sandwich wrappings, candy wrappers, dairy containers and lids, for portion packs and labels, and in general in the fast-food industry. Even more surprising is the manufacture of rigid and flexible containers for foods from biodegradable plastics that are produced from fermented sugars. These plastics are durable and water-resistant, but easily broken down by naturally occurring microorganisms to CO_2 and H_2O (Rice, 1991).

2.5.2 Controlling preservative migration and dough additives

Edible coatings controlling preservative migration from surface to food bulk could control surface microbial growth. Potassium sorbate is a common preservative that can be included in chitosan-, MC- and HPMC-based films. The permeability rate of the preservative can be associated with the relationships between coating film and coated food. Permeability rates follow the Arrhenius activation energy model. In a temperature range of 5–40°C, no breaking points in Arrhenius plots were observed, leading to the conclusion that no morphological changes within these films occurred. It seems that the solvent embedded within the film influenced activation energy, which in turn was independent of film composition. MC creates the best diffusion barrier. Such films had no channels or visible pores at a magnification of up to ×10 000 (Fakhrieh-Vojdani & Torres, 1990). The method of film preparation can influence its permeability to preservatives. Therefore, thermal properties of calcium alginate films were studied, together with the permeability of these films to sorbate and ascorbate. Use of the films to control diffusion of small-molecule food preservatives was also assessed. Calcium alginate films were prepared by either cooling the hot solutions (first method) or by pH-controlled release of calcium into the alginate solution (second method). A greater amount of energy was required to dissociate the cross-linked alginate by the second method. Differences in preparation also led to the latter film's higher permeability to potassium sorbate. The second method of preparation yielded films with estimated apparent activation energies for sodium ascorbate, potassium sorbate and ascorbic acid of 23.7, 24.1 and 36.2 kJ mol^{-1}, respectively (Wong *et al.*, 1996).

Sorbates in the range of 0.15–0.25% are permitted worldwide for the preservation of various food products. Sorbic acid and sorbates are effective in their non-dissociated state against fungi and certain bacteria. The use of coatings as carriers of sorbates improves their performance when applied to cut fruit or cheese analogues. This may be due to the slower diffusion of the preservatives into the food tissue, or to the presence of the preservative on the cut surface (Baldwin, 1999). The use of coatings to establish a surface pH that favors the active form of sorbic acid was demonstrated with an MC-palmitic acid film. Potassium sorbate permeability increased in parallel to an increase in water activity (Rico-Pena & Torres, 1991).

2.5.3 Composite bilayer, blends and biopolymer films

The components of edible films and coatings can be divided into three categories: hydrocolloids, lipids and composites. Composites contain both lipid (such as waxes, acylglycerols and fatty acids) and hydrocolloid components. A composite film can exist as a bilayer, in which one layer is the hydrocolloid and the other is the lipid, or as a conglomerate, where the lipid and hydrocolloid components are interspersed throughout the film (Donhowe & Fennema, 1992). The influence of plasticizer (PEG and glycerin) and three salts (potassium, calcium and magnesium chloride) on the WVP of biopolymer films containing various carrageenans was investigated. When no plasticizers or salts were included in films, their WVP was similar to those of wheat gluten and corn zein films, and was greatly affected by additives. The higher the KCl concentration for κ- and ι-carrageenan, the higher the observed WVP values. λ-Carrageenan films behaved similarly (Park *et al.*, 1996a).

A bilayer film composed of MC and palmitic acid, 3:1 (w/w), and an MC film (no additives), were cut in a circular shape to serve as moisture-impermeable barriers in a simulated sundae ice cream cone, positioned between the chocolate layer and the sugar cone. With the MC–palmitic acid films, no increase in moisture was detected in samples stored at −23°C and −12°C for 10 and 4 weeks, respectively (Rico-Pena & Torres, 1990). Other composite films were built of mixtures of sodium alginate and lipids smeared onto glass plates. Tensile strengths and WVP decreased with increasing stearic and palmitic acid concentrations. Paraffin was included in the formulation to create a paraffin-stearic acid-sodium alginate composite: these films had impressive moisture-barrier properties, i.e. their WVP was three times that of low-density polyethylene (Li *et al.*, 1996). Another cellulose-based material, HPMC (as one layer), was used to create a bilayer film, with stearic and palmitic acids: the lower the temperature, the higher the observed permeability values. The film functioned efficiently at relative humidity (RH) values of less than 90% (Kamper & Fennema, 1984). An MC-based film (one layer) and lipids (a beeswax layer deposited either from ethanolic solution or from a molten state), produced an interesting gum–lipid combination. Wax deposited from ethanolic solution yielded a film with higher WVP and OP values (Greener & Fennema, 1989). Another edible composite film made up of lipid and cellulose ethers was effective against moisture transport, even when the water activity on the low-humidity side of the film was relatively high. Lipid morphology is important in determining moisture resistance (Kester & Fennema, 1989). Chitosan–laurate composite film had low water permeability relative to other composites containing other fatty acids or esters. Their unique properties were attributed to the lipid morphological arrangement within the chitosan matrix (Wong *et al.*, 1992). In another report, removal of stearic acid crystals from an HPMC-stearic acid edible film changed the surface morphology and character, increasing permeability 10-fold (Hagenmaier & Shaw, 1990). The increased permeance of films based on cellulosic materials and lipids from bees wax, or a blend of bees wax and AGs, is probably caused by dehydration and swelling across the entire film thickness (Donhowe & Fennema, 1992). In the case of lipid–hydrocolloid films, a decrease in RH gradient contributes to a corresponding increase in water vapor permeance (Fennema *et al.*, 1994). Lipids are also useful in changing an edible film's resistance to CO_2 and O_2 transmission, improving surface gloss and handling properties (Greener & Fennema, 1992). There is also evidence that in laminated edible films (MC/corn zein-fatty acid films) the fatty acids are distributed from the corn zein layer into and through the MC layer (Park *et al.*, 1996b). In edible laminated films made from corn zein and MC, fatty acid migration increased with chain length and fatty acid concentration (Jang Woo Park, 1996).

Other plasticized blends prepared from high-amylose starch and citrus pectin formed unique films, whose mechanical properties were highly dependent on both plasticizer level and pectin:starch ratio, as well as on pectin–starch and pectin–plasticizer interactions (Coffin & Fishman, 1994). Different starches, cornstarch, waxy cornstarch, high-amylose cornstarch, wheat starch, potato and tapioca starch were also used to produce starch–PVA cast films. Those produced with high-amylose cornstarch exhibited the most consistent properties over the entire range of test conditions (Lawton, 1996).

Emulsion-production technology helped produce edible films of MC, stearic acid and PEG. A significant decrease in WVP through these films occurred with increasing stearic acid volume fractions (Sapru & Labuza, 1994). Thorough mixing is another way to combine soy isolate and sodium alginate, producing electrostatic complexes via interactions. These interactions maintained or improved soy protein solubility and emulsifying activity (Shih, 1994).

2.5.4 Biotechnological uses

Gums can be used for more unique non-traditional applications. For example, edible coatings based on HPMC were applied to the surface of green tomatoes in order to reduce the number and survival of *Salmonella montevideo* on the surface and in core tissue. A significant reduction in the number of viable bacteria on the surface was observed, whereas the reduction in the core tissue was less detectable. Addition of acetic, citric or sorbic acids at concentrations of up to 0.4% did not change this observation. Presence of ethanol (72–88%) was efficient for surface inactivation of bacteria. Changes in ethanol concentration did not change the internal inactivation of bacteria. In general, the coating was beneficial in retarding changes in color and texture, when fruit was stored at 20°C for up to 18 days (Zhuang *et al.*, 1996). Another study investigated coating physiologically mature tomatoes with edible sodium alginate films. A 2% alginate coating, containing 3–4% glycine and dried at 60–70°C, caused the fruit to turn brown-red at 9 days. Such films contributed more than 10 days to the fruit's shelf-life, and allowed it to maintain higher amounts of ascorbic acid, total acids, reducing agents and total sugar after 15 days of storage, compared to untreated controls at 5 days (Lui *et al.*, 1966). The source of gums from which edible films are produced is generally plants. Microorganisms manufacture fewer gums that can be used in foods. Polysaccharide produced by *Bacillus circulans* was compared with xanthan gum, locust bean gum (LBG), guar gum, sodium alginate and CMC. The polysaccharide exhibited high water-holding capacity, good emulsifying properties in oil and viscosity (1% solution) very similar to xanthan, and was unaffected by pH in the range of 5–9. Other properties of this polysaccharide solution were investigated, among them the influence of heat treatment, citric acid and glucose addition on viscosity. Addition of the polysaccharide to LBG and guar gum increased mixture viscosity, reaching maximum values at a ratio of 2:8 (v/v). Film prepared from this polysaccharide was tougher than those produced with CMC or guar gum (Isobe *et al.*, 1996).

Biotechnological uses of hydrocolloid films as coating agents can be found in the seed industry for traditional seed coatings. This approach has recently been pushed to a more refined level with the coatings of the first single cell and embryos (Fig. 2.5) with thin hydrocolloid films (Kampf *et al.*, 2000a, b). Coating is different from entrapment and immobilization in that the coating around the cell is thinner, comprising only a small fraction of the cell or embryo's diameter. The coating of very large cells, such as oocytes, could also be useful to the food industry. Such foods could include caviar or other fish eggs, and a coating could provide thermal stabilization to these delicate objects; better heat endurance could be achieved via chemical interactions between the coating and the egg membrane (Kampf *et al.*, 2000a, b; Nussinovitch & Kampf, 2001).

2.5.5 Coated paper

Carrageenans are regularly used for milk and meat texturized products, and for use as thickeners and stabilizers in foods, besides other uses. The resistance of paper to lipid penetration can be changed and potentially controlled by the addition of κ-, ι-or λ-carrageenan to produce biodegradable packaging material for foods. Both carrageenan-coated and non-coated films are resistant to lipid penetration; however, κ-carrageenan-coated papers of greater than 4 kg per ream exhibit adequate lipid-barrier properties for use in lipid-containing food products, and resistance was found to increase with film thickness (Rhim *et al.*, 1996).

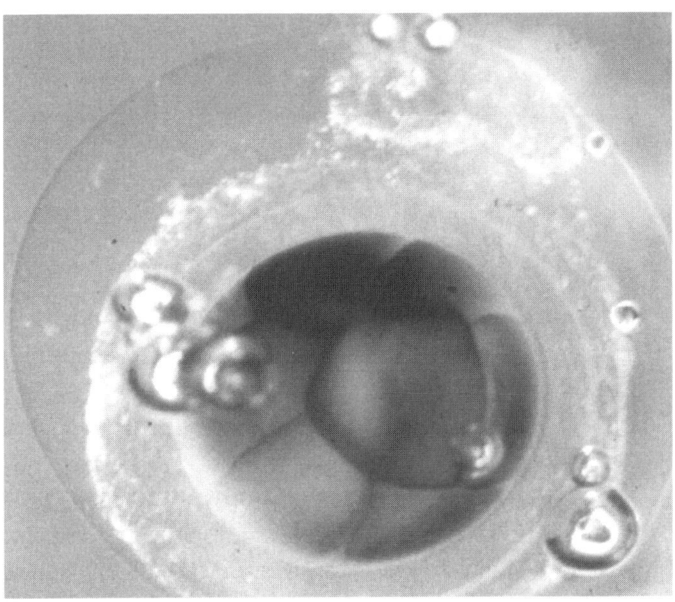

Fig. 2.5 *Xenopus laevis* embryo coated by alginate. (Courtesy of Dr. N. Kampf.)

2.5.6 Special hydrocolloid disturbances in wax-based coatings

The Florida citrus industry was the first to realize the value of coating by dip application for fresh citrus fruit. First, paraffin-based coatings were used to control weight loss. Later, resin–water solution coatings containing shellac and/or alkali-soluble resins, adjuvants, plasticizers and emulsifiers were developed to satisfy the need for longer shelf-life and better shine, in an attempt to improve coated fruit salability (Chen & Nussinovitch, 2000a). Wax coatings were developed to mimic the natural coating of fruits and vegetables. However, these coatings inhibited respiratory gas exchange to such an extent that fermentation was induced. As a result, ethanol buildup was detected along with that of other volatiles, coinciding with a fermented and bitter taste. At high levels, these volatiles are considered off-flavors and they reduce fruit quality (Sinclair, 1961). Xanthan gum was introduced into a traditional, wax-based coating formulation for easy peelers. Xanthan was chosen because of its nature as a non-gelling agent and the fact that coatings based solely on this gum have the highest gloss relative to that of films produced with other non-gelling agents. Xanthan created disturbances in the ordered, regular structure of the traditional wax coating, as observed by EM. As a result of this imperfect coating, fruit respiration was less disturbed and less ethanol and acetaldehyde were detected by gas chromatography in the fruit as compared to fruits that were coated by the same method using a commercial formulation. In addition, less off-flavors were detected by sensory evaluation of juice extracted from the fruit coated with the wax–xanthan coating. Xanthan therefore seems to be a beneficial addition to traditional wax coatings, producing tastier fruits and juices (Chen & Nussinovitch, 2000a). The same authors also tried introducing LBG or guar gum into traditional wax formulations of two easy-peeler citrus fruit cultivars: Nova and Michal (Fig. 2.6). Their performance was analyzed and compared to inclusion of the non-gelling xanthan gum in similar formulations. LBG and guar gum have been observed to reduce weight loss in fruits during respiration, similar to wax-based coatings without gum. No disadvantage relative to a commercial coating was observed in terms of ethanol and acetaldehyde buildup. Juice with the best taste quality was produced from fruits coated with xanthan- and LBG-wax coatings. Galactomannans did not change the gloss of the coated easy peelers. LBG, similar to xanthan, showed superior performance as a coating additive. Inclusion of such gums within the wax contributed to a more chaotic, less-ordered structure, highly desirable in terms of improving fruit respiration (Chen & Nussinovitch, 2000b, 2001).

2.6 Film-application techniques

Films can be applied by different methods, namely dipping, spraying or coating. Dipping of a food product in one or several baths (if cross-linking of the gum is required) is followed by draining and drying. This method, aside from its usefulness for coating foods with a dried gum film, can be used to apply wax coatings to agricultural products (Krochta *et al.*, 1994; Nussinovitch, 1997). In many laboratories, the casting technique is used to produce films of nearly constant thickness. The gel or

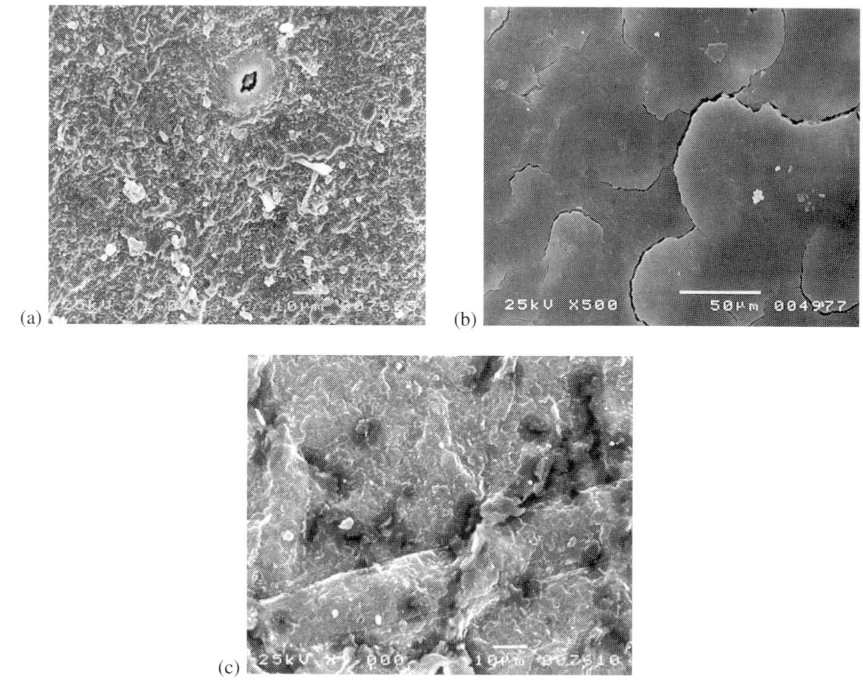

Fig. 2.6 SEM micrographs of wax–hydrocolloid-based coatings of citrus fruit (easy-peelers). (a) No coating. (b) Commercial coating. (c) Carnauba–LBG coating. (Courtesy of Mr. S. Chen.)

hydrocolloid solution is poured into a *sandwich* prepared from glass plates, assembled with screw clamps and aligned using the casting stand's alignment slot and an alignment card. After the gels are cast, they are left to equilibrate under high humidity at room temperature. The films are then dried with warm air until a predetermined amount of moisture remains within them. Specimens are then prepared with a dumb-bell shaped cutter for tensile tests (Nussinovitch, 1997). Brushes, falling-film enrobing technique, panning or rollers can also be used to apply films to the surfaces of fruits and vegetables (Guilbert, 1986).

2.7 Physical methods and relevant parameters before, during and after coating application

2.7.1 Parameters to be considered before and during coating

The coating of a vegetable skin with a gelling agent solution that requires later cross-linking by cations involves a process that could hypothetically be characterized by four successive steps and time scales. Immersion of the fresh produce in a gum coating solution (not including the cationic cross-linking agent) can take between ~15 s and 2 min to yield complete coating of the object. The duration depends on the wettability, the concentration and the viscosity of the hydrocolloid solution, as well as on the roughness of the surface and the possible penetration of the coating liquid into the biological specimen (Hershko *et al.*, 1996). The second immersion of the gum-coated commodity, to cross-link the gum (i.e. alginate, LMP, gellan, etc.), takes between 30 s and ~2 min, depending upon the concentration and temperature of the cross-linking agent and the gum solution, the thickness of the coating gum solution and the geometric complexity of the coated object. The third stage of coating, namely continuous strengthening of the gel coating layer, is usually accomplished at high relative humidity. This is done to avoid dehydration and to achieve different gel strengths, depending on duration. This stage can last between 20 min and ~2 h and also depends on the temperature and concentrations of hydrocolloids and other compounds involved. The fourth stage, namely drying, can result in different dry-film textures and structures, depending on its duration. When the coated object is subjected to drying, after a short or long gelation process, texture and structure vary according to the time required for the gelled film layer to lose its inherent moisture. The properties of the dried films also depend on the properties of the drying equipment, the thickness of the coating film and its composition, the T_g of the polymer, and the final mechanical properties of the *wet* film, obtained before the drying process takes place (Hershko & Nussinovitch, 1998a, b). The coating process involves wetting of the produce to be coated by the coating gum solution, possible penetration of the solution into the skin (Hershko *et al.*, 1996), followed by possible adhesion between these two commodities. The wetting stage (spreadability) is, of course, the shortest and most important one because, if the suitability of the compound used for spreading and the object to be coated are ideal, the time interval necessary for such an operation is minimal, or, in other words, spreadability is spontaneous (Mittal, 1983). However, since it is almost impossible to find hydrocolloid solutions or their combinations that are perfectly suited to the surface properties of the object, such as surface tension and polarity, the user should seek the best possible combination to approximate such compatibility. To achieve a successful tailored coating that adheres to a fruit or vegetable skin, the inter-facial tension between the coating solution and the vegetable surface should be estimated. The inter-facial tension γ_{SL} depends on the surface tension of the solid skin, γ_{SV}, the surface tension of the adherent (e.g. the coating solution), γ_{LV}, and the contact angle, θ, of a specific coating solution on a solid surface in accordance with the Young equation:

$$\gamma_{LV} \cos\theta = \gamma_{SV} - \gamma_{SL} \tag{1}$$

in which γ denotes surface tension and subscripts L, V, and S denote liquid, vapor and solid, respectively (Hershko & Nussinovitch, 1998a).

The adhesion of polymeric coatings has been studied for the metal industry (Mittal, 1983). Many reports have dealt with the relationships between ineffective adhesion and rougher surfaces. However, there are well-established examples of surface roughness enhancing adhesion. Early researchers considered adhesion to be of two main types: specific and mechanical. The former led to the development of the adsorption theory. Mechanical adhesion was hypothesized to occur when a liquid set in the pores and cracks of a substrate, providing a mechanical key. When mechanical keying occurs, adhesion is expected to increase with substrate roughness. The effect of surface topography on measured adhesive-bond strength, as detailed in published reports, is complex and to some extent contradictory. A roughened surface can lead to incomplete wetting and reduce the area of contact, and voids at the interface may increase stress which, with brittle adhesives, could weaken the joint. The effect of surface roughness on coatings has been studied in food and agricultural commodities (Hershko *et al.*, 1996), but data on the roughness of fruit and vegetable surfaces is rarely found (Ward & Nussinovitch, 1996). More effective coatings could be developed by exploring properties of fruit and vegetable surfaces, such as roughness. Surface topographies of onion and garlic skins before coating were studied and a method of estimating these properties was developed (Hershko *et al.*, 1998). Atomic force microscopy (AFM) was used to estimate the surface area. Image-processing and Arc/Info software were used to interpret the data. The calculated ratio between apparent and measured surfaces (roughness factor) ranged from 1.11 to 1.15 for untreated and chloroform-treated onion skin, respectively. For garlic, higher values were detected for the untreated skin. The higher the roughness factor, when the coating solutions are easily spread on the fruit or vegetable surfaces, the better the adhesion between the coating and the skin. A knowledge of true surface areas can help to better estimate the required coating-solution volumes (Hershko *et al.*, 1998). In all tested specimens taken from garlic or onion the roughness factor, r, was higher than 1.0, in accordance with any such surface in practice. Gum solutions usually used for coating include CMC, cellulose ethers, carrageenan, alginate and pectin, as well as many other natural and synthetic water-soluble polymers. The contact angles of such materials were measured on garlic and onion skins (Hershko & Nussinovitch, 1998c), yielding values of $<90^\circ$ in all cases. Combined with the observed $r>1.0$, this indicates enhanced spreading of the coating solution on the vegetable or fruit surface, better contact between solution and surface, resulting in good adhesion. These results indicate that changes in fruit-surface roughness by delicate mechanical or chemical treatments could be beneficial in terms of more effective coating adhesion (Hershko *et al.*, 1998). Three other reports emphasize and demonstrate that if wetting gum solutions include ingredients similar to those naturally incorporated on the surface of the coated commodity, better chemical similarity can be achieved, and it is most important for the success and adhesion of the coating. For coating purposes, foreknowledge of the properties of the surface to be coated and the gum to be used is desirable (Hershko & Nussinovitch, 1998a, b, c).

The surface tension of gelling and inducing solutions, and their contact angles on fruit and vegetable surfaces, can be studied with surface-tension instruments (maximum adhesion requires a contact angle of 0°). It is important to note that 'if the coating does not spread spontaneously over the substrate surface, so that there is inter-molecular contact between the substrate surface and the coating, there cannot be interactions and hence no contribution to adhesion' (Wicks *et al.*, 1994). The main aim of those dealing with coatings is to reduce the surface tensions of the coating solutions, to adjust them to the lower surface tension of the fruit or vegetable surface, and to improve wettability (Wu, 1973). Elimination of dewetting is presumably much more important from a commercial point of view, especially when the tested combinations adhere to the surface of the commodity to be coated (Hershko & Nussinovitch, 1998a). To estimate solid surface tension, γ_{SV}, critical surface tension should first be calculated. To comply with this requirement, the critical surface tension of the object to be coated should be derived from Zisman plots followed by extrapolation. Zisman plots are obtained by calculating the cosine of each of the measured contact angles of the tested solutions on the commodity skin and plotting it against the already-known surface tension values of the solvents and binary mixtures used for the study (Fig. 2.7). If there is no information on the surface tension of the vegetative skin, it is recommended to estimate those solid surface tension values by another method, based on visual observation of the spreading of liquids with known surface tension and with partial components of the surface tension, i.e. dispersion and polarity, upon solid surfaces. Thus an approximate value for γ_c, the critical surface tension of a solid, can also be obtained and compared with results derived from Zisman plots (Hershko & Nussinovitch, 1998a). Since the reliability of the common methods for the estimation of surface tensions of viscous hydrocolloid solutions has been questioned, an examination of the results using the

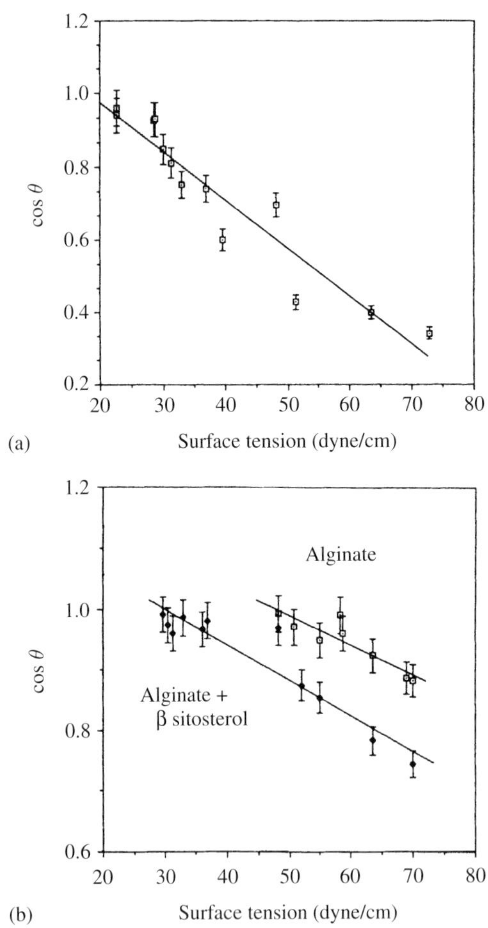

Fig. 2.7 (a) Zisman plot for garlic acid. (b) Zisman plot for dry alginate and alginate–sitosterol films. (From Hershko & Nussinovitch, 1998a, with permission from American Chemical Society.)

pendant-drop method designed especially for viscous solutions is recommended. In such cases, γ_L is calculated in accordance with

$$\gamma_L = \Delta\rho g d_e^2 / H \tag{2}$$

in which $\Delta\rho$ is the difference in density between air and the tested solution, g is the acceleration constant and $1/H$ is the quantity achieved from numerical tabulation of $S\ (d_S/d_e)$ against $1/H$, which was first obtained by Niederhauser and Bartell (1950).

The interfacial tension γ_{SL} is an important factor in wettability. If it is greater than the difference between the surface tension of the solid and the surface tension of the solution, no wetting will occur. In other words, the spreading coefficient G needs to be positive to obtain complete or optimum wettability:

$$G_{SL} = \gamma_S - \gamma_L - \gamma_{SL} \tag{3}$$

The greater the positive G value, the greater the free energy loss when a specific liquid spreads on a specific solid and the more stable the combination of the two components to be adhered (Gans, 1966). Similar calculations were performed by us for garlic skin and its hydrocolloid coatings (Hershko & Nussinovitch, 1998a). It was concluded that tailor-made hydrocolloid coatings for different vegetative materials can only be achieved by further exploring the chemical and physical properties of the coating solutions and the coated objects.

During coating, interactions between the coating materials and the surface of the object should be considered. The surface of food can be more or less smooth. Thus its roughness should be studied. Roughness of a surface can be determined by a roughness tester, and by analyzing AFM *maps*. Parameters that can be studied are R_t, the distance between the highest peak and the deepest valley of the roughness profile within an evaluated length of tested surface, and R_a, the arithmetic mean of the absolute values of the roughness profile deviation from the center line within the evaluation length (Nussinovitch, 1997).

In a study with mushroom tissue (Fig. 2.8), the amount of penetrating gum solution depended mainly on the gum's physical and chemical properties. Its viscosity, surface tension, flow pattern, hydrophobic–hydrophilic nature, osmotic pressure and other factors related to these properties are important in trying to explain the complexity of the simple coating of a porous vegetable surface by a gum solution (Hershko & Nussinovitch, 1998b). From previous studies (Hershko *et al.*, 1996), it is well known that it is easier to follow traces of migrating minerals within the coated object than to detect absorbed hydrocolloids within the vegetable tissue. Here X-ray analysis and inductively coupled plasma (ICP) could be very beneficial.

2.7.2 *Parameters to be considered after coating*

Permeability of gum films can be measured by several methods (Anon, 1975; Karel, 1975; Stern *et al.*, 1964). Many apparatuses can be found on the market for determining the permeability of films to gases (Landrock & Proctor, 1952; Quast & Karel, 1972).

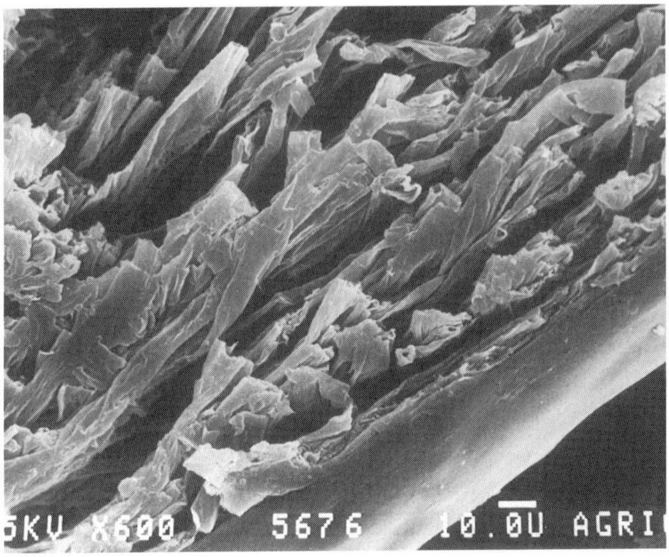

Fig. 2.8 Cross section of alginate-coated mushroom. Coating on the right side. (Courtesy of Dr. V. Hershko.)

Rates of water vapor transmission through dried coatings can be determined by ASTM E393-83 (standard test methods for water vapor transmission of materials) (ASTM, 1983).

Water vapor transmission of edible films and coatings from proteins, polysaccharides and lipids is commonly measured using modifications of the ASTM E96 standard method (cup method). A stagnant air layer exists between the underside of the test film mounted on the cup and the surface of the desiccant, a saturated salt solution of distilled water in the cup (Gennadios *et al.*, 1994). A continuous gravimetric method has been proposed to accurately measure the WVP of edible films. In a typical study, MC and paraffin wax were used to produce a film. Film permeability was dependent on polarity, homogeneity of dispersed material in the film and its structure, thickness and water vapor transmission, which in turn increased with hydrophilicity and heterogeneity. Laminating the film with wax changed its barrier efficiency (Debeaufort *et al.*, 1993). Other films produced from furcellaran were studied at 22, 35 and 50°C in the vapor activity range of 0.1–0.96. Methods for determining film properties included sorption isotherm analysis using the Zimm–Lundberg procedure and evaluation of activation energy values (Ptitchkina & Chalykh, 1994). For laminated MC-corn zein fatty acid films, an evaluation of quality and product suitability included studying WVP, tensile strength and percent elongation of the film under mechanical testing (Park *et al.*, 1994). Carrageenan-based edible films were investigated for their thickness, RH, water vapor transmission rate, activation energy of this rate and its dependence on film thickness (Jong Whan Rhim *et al.*, 1996). In addition, the force required to peel the coating from the coated object can be determined by peel test. This method is used to estimate the degree of adhesion of the film to the object. The procedure requires that the coating be peeled at 90° from the substrate, and that the adhesion strength be estimated by the force per unit width neessary to peel the coating (Nussinovitch, 1997).

2.8 Plasticizers

Many wax-type coatings are made with natural or synthetic waxes and fatty acids, oils, wood rosin, shellac emulsifiers, anti-foam agents, surfactants, preservatives and plasticizers. Edible films can aid in food preservation. Plasticizers serve as the necessary additives in films produced by blends of pectins or alginates with milk proteins. The addition plays a role in changing the film's properties, such as water vapor barrier quality and tensile strength. With a proper choice of plasticizer, WVPs of films with a basic hydrophilic nature can be increased without negatively affecting tensile properties. In such cases, the incorporation of whole milk, sodium caseinate, non-fat dry milk or whey can be considered. Lower WVP values were observed for sodium alginate than for high- or low-methoxy pectins. An efficient plasticizer such as sodium lactate, at 50% or higher, conferred an elongation value greater than 13%. Sorbitol is an even better plasticizer where WVP is concerned, but its negative contribution to the increase in brittleness cannot be ignored (Parris *et al.*, 1995). The reader is also referred to Section 2.4.7 for a discussion of PEG inclusion in cellulose films.

2.9 Drying of films

In addition to decreasing the film's water content and changing its structure in order to influence its barrier properties, drying is often used to help adhere the film to the product, or at least to bring the coating film and the coated object into maximal contact. Aqueous film solutions of MC (including 0–75% ethanol, in 25% increments), were dried either at 100°C for 35 min, 80°C for 1 h, 50°C for 1.5 h or at room temperature overnight. At a water-to-ethanol ratio of 3:1, MC films exhibited smaller permeabilities to oxygen and water vapor, and greater crystallinity, tensile strength and percent elongation at higher drying temperatures. Higher ethanol concentrations can completely eliminate hydration and at 25%, enhance inter-molecular hydrogen bonding of MC. This, in turn, along with the hydrophobic association which induces crystallization, reinforces the film to produce enhanced resistance to water vapor and oxygen (Donhowe & Fennema, 1993).

Drying as an essential step in coating can be evaluated by several methods (Bowser & Wilhelm, 1995). New techniques for drying thin films of colloidal slurries and solutions have been introduced. Their advantages lie in faster dehydration rate and better quality of the formed film due to the low temperature at which the film solidifies (Bowser & Wilhelm, 1996). Heat, mass transfer and shrinkage of the proposed thin-film drying system are described by a mathematical model. Physical properties of starch, including moisture diffusivity, heat of sorption, modulus of elasticity, bulk shrinkage, water activity and thermal diffusivity, were taken from the literature and used in the model (Bowser, 1995).

2.10 Gloss properties of hydrocolloid films

The production of films involves the drying of gel gum solutions. Moisture moves through the system to the surface by different mechanisms (Watson & Harper, 1983). At the surface, the moisture evaporates and is carried away in a circulating air stream. This process results in the sedimentation of particles from a gum solution and the subsequent formation of a thin, dry film layer. In gels, the drying process results in catastrophic collapse of the hydrocolloid network and formation of a film. The collapse of polyacrylamide gel networks was observed upon immersion in acetone or certain salt solutions, resulting in shrunken products (Tanaka, 1981).

A very small number of studies on the gloss properties of hydrocolloid films and their dependence on different factors can be found in the literature. In order to study the gloss of hydrocolloid films and its dependence on different parameters (e.g. gum concentration), these moieties were produced from xanthan, alginate, agar or agarose. Gellan films having gum concentrations of 0.25–3.0% and gelatin films with an initial water-soluble polymer concentration of 4–15% have been manufactured (Ward & Nussinovitch, 1996, 1997). The film thickness increased with hydrocolloid concentration in the gel or gum solution. This may correspond to an increase in the dry solid content of the films. For example, films produced from agar gels increased in thickness from 0.007 ± 0.003 mm at 0.5% agar to 0.091 ± 0.041 mm at 3% agar (i.e. a sixfold increase in agar concentration led to a 13-fold increase in thickness). Gloss measurements of these hydrocolloid films were made at a 60° angle using a flat-surface glossmeter. High standard deviations were obtained for gloss measurements due to small, local deviations in the surface profile of the dried films. This is based on the fact that specular reflectance of a surface is a sensitive function of its roughness. Such deviations in surface profile were manifested in roughness measurements and observed in SEM studies. The mean values of gloss for the dried gels indicated that some trends may exist. Gloss of films produced from agar decreased from a gloss value of 38.6 at a 0.5% gel concentration to 20.6 at a 3% concentration. Linear regression analysis provided a high significant correlation coefficient between mean gloss and agar concentration (Ward & Nussinovitch, 1996, 1997).

2.11 References

Adachi, S., S. Segawa & R. Matsuno (1995) Surface property of dehydrated protein or polysaccharide film and mass transfer through the film. *Report of the Soy Protein Research Committee*, **16**, 109–114 (in Japanese).

Alikonis, J.J. (1979) *Candy Technology*. AVI Publishing Company, Inc., Westport, CT.

Andres, C. (1984) Natural edible coating has excellent moisture and grease barrier properties. *Food Processing*, **45**(12), 48.

Anon (1975) Instructions for the Oxtran-100 Oxygen Permeability Tester. Mocon Instruments, Inc., Minneapolis, MN.

Anon (1977) Edible and water soluble films. *Food Engineering*, **August**, 49–50.

Anon (1980) Manufacture of Koenigsberger marzipan. Traditional and rationalized methods. *Zucker und Suesswarenwirtschaft*, **33**(7/8), 252, 255.

Anon (1993) Food gums stick it out in dry mix glazes. *Prepared Foods*, **162**(1), 53.

Anon (1995) Use of tapioca dextrins for dragee confectionery. *Suesswaren*, **39**(4), 20–22.

Anon (1997) Edible films solve problems. *Food Technology*, **51**(2), 60.

Arai, L., Y. Kiunmaki & T. Fujita (1968) Toxicity of chitosan. *Bull. Takai Reg. Fish Res. Lab*, **56**, 89.

Arnold, F.W. (1963) Gum Arabic coatings for nut products. United States Patent 3,383,220.

Arvanitoyannis, I., E. Psomiadou, A. Nakayama, S. Aiba & N. Yamamoto (1997) Edible films made from gelatin, soluble starch and polyols. *Food Chemistry*, **60**(4), 593–604.

Ayranci, E. (1996) Moisture sorption of cellulose-based edible films. *Nahrung*, **40**(5), 274–276.

Baldwin, E.A. (1994) Edible coatings for fresh fruits and vegetables: past, present and future. In: (Krochta, J., E. Baldwin & M. Nisperos-Carriedo, Eds.), *Edible Coatings and Films to Improve Food Quality*. Technomic Publ. Co., Basel, Switzerland.

Baldwin, E.A. (1999) Surface treatments and edible coatings in food preservation. In: (Shafiur Rahman, M., Ed.), *Handbook of Food Preservation*. Marcel Dekker, Inc., New York and Basel, pp. 577–609.

Baldwin, E.A., M.O. Nisperos-Carriedo & R.A. Baker (1995) Use of edible coatings to preserve quality of lightly processed products. *Critical Reviews in Food Science and Nutrition*, **35**(6), 509–524.

Balke, W. (1977) Coating gum. *Manufacturing Confectioner*, **57**(11), 63–64.

Banks, N.H. (1984) Studies of the banana fruit surface in relation to the effects of TAL Prolong coating on gaseous exchange. *Sci. Hort.*, **24**, 279–286.

Barkai-Golan, R. (1990) Postharvest disease suppression by atmospheric modifications. In: (Calderon, M. & R. Barkai-Golan, Eds.), *Food Preservation by Modified Atmospheres*. CRC Press, Boca Raton, FL, pp. 238–265.

Bauer, C.D. & G.L. Neuser (1969) Edible meat coating composition. United States Patent 3,483,004.

Ben-Yehoshua, S. (1966) Some effects of plastic skin coating on banana fruit. *Trop. Agric. Trin.*, **43**, 219–232.

Blake, J.R. (1966) Some effects of paraffin wax emulsions of bananas. Queensland. *J. Agric. Animal Sci.*, **23**, 49–56.

Bose, A.N. & G. Basu (1954) Studies on the use of coating for extension of storage life of fresh Fajli mango. *Food Research*, **19**, 424–428.

Bowser, T.J. (1995) Thin film drying on a permeable surface. *Dissertation Abstracts International*, **55**(7), 2868.

Bowser, T.J. & L.R. Wilhelm (1995) Modeling simultaneous shrinkage and heat and mass transfer of thin nonporous film during drying. *J. Food Sci.*, **60**(4), 753–757.

Bowser, T.J. & L.R. Wilhelm (1996) A water vapor permeable drying surface for thin films. *Transactions of the ASAE*, **39**(2), 617–623.

Brake, N.C. & O.R. Fennema (1993) Edible coatings to inhibit lipid migration in a confectionary product. *J. Food Sci.*, **58**(6), 1422–1425.

Brown, E. (1984) Efficacy of citrus postharvest fungicides applied in water or resin solution water wax. *Plant Dis.*, **68**, 415–418.

Bull, A. (1996) Special starches facilitate production. *Zucker und Suesswaren Wirtschaft*, **49**(9), 422–423.

Butler, R.W. & E.D. Klug (1980) Hydroxypropylcellulose. In: (Davidson, R.L., Ed.), *Handbook of Water-Soluble Gums and Resins*. McGraw-Hill Book Company, NY, pp. 13-1–13-17.

Butler, B.L., P.J. Vergano, F.R. Testin, J.M. Bunn & J.L. Wiles (1996) Mechanical and barrier properties of edible chitosan films as affected by composition and storage. *J. Food Sci.*, **61**(5), 953–955.

Byrne, M. (1982) The dry roast revolution. *Food Manufacture*, **57**(10), 55, 57.

Chattopadhyay, S., R.S. Singhal & P.R. Kulkarni (1997) Optimization of conditions of synthesis of oxidised starch from corn and amaranth for use in film-forming applications. *Carbohydrate Polymers*, **34**(4), 203–212.

Chen, M.C., G.H. Yeh & B.H. Chiang (1996) Antimicrobial and physiochemical properties of methylcellulose and chitosan films containing a preservative. *J. Food Proc. Preserv.*, **20**(5), 379–390.

Chen, S. & A. Nussinovitch (2000a) The role of xanthan gum in traditional coatings of easy peelers. *Food Hydrocolloids*, **14**, 319–326.

Chen, S. & A. Nussinovitch (2000b) Galactomannans in disturbances of structured wax-hydrocolloid-based coatings of citrus fruit (easy-peelers). *Food Hydrocolloids*, **14**, 561–568.

Chen, S. & A. Nussinovitch (2001) Permeability and roughness of wax-hydrocolloid coatings and their limitations in determining citrus fruit overall quality. *Food Hydrocolloids*, **15**, 127–137.

Coffin, D.R. & M.L. Fishman (1994) Mechanical properties of pectin-starch films. *Polymers from Agricultural Coproducts*. American Chemical Society Meeting, American Chemical Society, Washington DC, pp. 82–91.

Cohen, E., Y. Shalom & I. Rosenberger (1990) Postharvest ethanol buildup and off-flavor in "Murcott" Tangerine fruits. *J. Amer. Soc. Hort. Sci.*, **115**, 775–778.

Cole, M.S. (1965) Method for coating dehydrated food. United States Patent 3,479,191.

Cuq, B., N. Gontard & S. Guilbert (1997) Packaging materials based on natural polymers. *Industries Alimentaires et Agricoles*, **114**(3), 110–116.

Curtis, G.J. (1988) Some experiments with edible coatings on the long-term storage of citrus fruits. *Proc. 6th International Citrus Congress*, **3**, pp. 1514–1520.

Dalal, V.B., P. Thomas, N. Nagaraja, G.R. Shah & B.C. Amla (1970) Effect of wax coating on bananas of varying maturity. *Indian Food Packer*, **24**, 36–39.

Daniels, R. (1973) *Edible Coatings and Soluble Packaging*. Park Ridge, Noyes Data Corp, NJ.

Davidson, R.L. (1980) *Handbook of Water-Soluble Gums and Resins*. McGraw-Hill Book Company, New York.

Debeaufort, F. & A. Voilley (1995) Effect of surfactants and drying rate on barrier properties of emulsified edible films. *International Journal of Food Science & Technology*, **30**(2), 183–190.

Debeaufort, F., M. Martin Polo & A. Voilley (1993) Polarity homogeneity and structure after water vapor permeability of model edible films. *J. Food Sci.*, **58**(2), 426–429, 434.

Debeaufort, F., J.A. Quezada-Gallo & A. Voilley (1998) Edible films and coatings: tomorrow's packaging: a review. *Critical Reviews in Food Science*, **38**(4), 299–313.

Donhowe, I.G. & O. Fennema (1992) The effect of relative humidity gradient on water vapor permeance of lipid and lipid-hydrocolloid bilayer films. *Journal of the American Oil Chemists' Society*, **69**(11), 1081–1087.

Donhowe, I.G. & O. Fennema (1993) The effects of solution composition and drying temperature on crystallinity, permeability and mechanical properties of MC films. *J. Food Proc. Preservation*, **17**(4), 231–246.

Eaks, I.L. & W.A. Ludi (1960) Effects of temperature, washing and waxing on the composition of the internal atmosphere of orange fruits. *Proc. Amer. Hort. Soc.*, **76**, 220–228.

Earle, R.D. & D.H. McKee (1976) Process for treating fresh meats. United States Patent 3,991,218.

Earle, R.D. & C.E. Snyder (1966) Method of preparing frozen seafood. United States Patent 3,255,021.

El Ghaouth, A., A.J. Arul, R. Ponnampalan & M. Boulet (1991) Chitosan coating effect on storability and quality of fresh strawberries. *J. Food Sci.*, **56**, 1618.

Eskins, K., G.F. Fanta, F.C. Felker & F.L. Baker (1996) Ultrastructural studies on microencapsulated oil droplets in aqueous gels and dried films of a new starch-oil composite. *Carbohydrate Polymers*, **29**(3), 233–239.

Fakhrieh-Vojdani & J.A. Torres (1990) Potassium sorbate permeability of polysaccharide films: chitosan, methylcellulose and hydroxypropyl methylcellulose. *J. Food Proc. Eng.*, **12**(1), 33–48.

Fellows, P.J. (1990) *Food Processing Technology Principles and Practice.* Ellis Horwood, New York.

Fennema, O., I.G. Donhowe & J.J. Kester (1994) Lipid type and location of the relative humidity gradient influence on the barrier properties of lipids to water vapor. *J. Food Eng.*, **22**(1–4), 225–239.

Fisher, D.V. & J.E. Britton (1940) Apple waxing experiments. *Sci. Agric.*, **21**, 70–79.

Gans, D.M. (1966) Wetting, spreading and contact angles. *J. Paint Technol.*, **38**(497), 322–326.

Ganz, A.J. (1977) Cellulose hydrocolloids. In: (Graham, H., Ed.), *Food Colloids*. AVI Press, Westport, CT, pp. 382–417.

Gates, J.C. (1981) *Basic Foods*. Holt, Reinhart and Winston, New York, pp. 99–153.

Gennadios, A., C.L. Weller & C.H. Gooding (1994) Measurement errors in water vapor permeability of highly permeable hydrophilic edible films. *J. Food Eng.*, **21**(4), 395–409.

Gennadios, A., M.A. Hanna & L.B. Kurth (1997) Application of edible coatings on meats, poultry and seafoods: a review. *Lebensm.-Wiss. u.-Technol.*, **30**, 337–350.

Glascow, G.U. & A.J. Kraght (1970) Gloss coating for fruits. United States Patent 3,488,200.

Greener, I.K. & O. Fennema (1989) Barrier properties and surface characteristics of edible, bilayer films. *J. Food Sci.*, **54**(6), 1393–1399.

Greener, I. & O. Fennema (1992) Lipid-based edible films and coatings. *Lipid Technol.*, **4**(2), 34–38.

Greener, I. & O. Fennema (1993) The effects of plasticizers on crystallinity, permeability and mechanical properties of MC films. *J. Food Process Preservation*, **17**(4), 247–257.

Guilbert, S. (1986) Technology and application of edible protective films. In: (Mathlouthi, M., Ed.), *Food Packaging and Preservation Theory and Practice*. Elsevier Applied Science Publishing Co., London, England.

Hagenmaier, R.D. & P.E. Shaw (1990) Moisture permeability of edible films made with fatty acid (hydroxypropyl) methylcellulose. *J. Agric. Food Chem.*, **38**(9), 1799–1803.

Hannigan, K. (1983) Nutrients 'locked into' food: edible coating. *Food Engineering*, **55**(1), 59.

Hardenburg, R.E. (1967) Wax and related coatings for horticultural products. A bibliography. *Agr. Res. Bull.* 51–15, U.S. Dept. of Agric., Washington, DC.

Hercules, Inc. (1978) *Cellulose Gum – Chemical and Physical Properties*. Hercules Inc., Wilmington, DE.

Hershko, V. & A. Nussinovitch (1996) Gellan and alginate vegetable coatings. *Carbohydrate Polymers*, **30**, 185–192.

Hershko, V. & A. Nussinovitch (1998a) The behavior of hydrocolloid coatings on vegetative materials. *Biotechnol. Prog.*, **14**, 756–765.

Hershko, V. & A. Nussinovitch (1998b) Relationships between hydrocolloid coating and mushroom structure. *J. Agric. Food Chem.*, **46**(8), 2988–2997.

Hershko, V. & A. Nussinovitch (1998c) Physical properties of alginate-coated onion (*Allium cepa*) skin. *Food Hydrocolloids*, **12**, 195–202.

Hershko, V., E. Klein & A. Nussinovitch (1996) Relationships between edible coatings and garlic skin. *J. Food Sci.*, **61**(4), 769–777.

Hershko, V., D. Weisman & A. Nussinovitch (1998) Method for studying surface topography and roughness of onion and garlic skins for coating purposes. *J. Food Sci.*, **63**(2), 317–321.

Hilgeland, P.H. (1964) Apparatus for coating. British Patent 967,501.

Hill, S.E., D.A. Ledward & J.R. Mitchell (1998) *Functional Properties of Food Macromolecules*. An Aspen Publication, Gaithersburg, MD, p. 188.

Hitz, C.W. & I.C. Haut (1938) Effect of certain waxing treatments at time of harvest upon the subsequent storage quality of 'Grimes Golden' and 'Golden Delicious' apples. *Proc. Amer. Soc. Hort. Sci.*, **36**, 440–447.

Hoagland, P.D. (1996) Films from pectin, chitosan and starch. *Macromolecular Interactions in Food Technology*. American Chemical Society, Washington DC, pp. 145–154.

Hoagland, P.D. & N. Parris (1996) Chitosan/pectin laminated films. *J. Agric. Food Chem.*, **44**(7), 1915–1919.

Hyun J. Park & M.S. Chinnan (1995) Gas and water vapor barrier properties of edible films from protein and cellulosic materials. *J. Food Engineering*, **25**(4), 497–507.

IFT (1991) New from Mitsubishi. Annual Meeting and Food Expl. Program and Exhibit Directory, Dallas Convention Center, 1–5 June, 1991, Chicago, IL.

Ijichi, K.E. (1978) *Evaluation of an alginate coating during frozen storage of red snapper and silver salmon*. MSc thesis, University of California-Davis.

Isobe, Y., Y. Toyama, M. Minamori, K. Yokoigawa, K. Endo, F. Kawai & H. Kawai (1996) Comparison of physicochemical properties of a novel polysaccharide produced by *Bacillus circulans* with commercial polysaccharides. *J. Japanese Soc. Food Sci. Technol.*, **43**(5), 634–641.

Jang Woo Park (1996) Development and property enhancement of bi-layer edible films produced from corn zein and methylcellulose. *Dissertation Abstracts International*, -B; **56**(9), 4663.

Jeanes, J. (1976) Dextrans and pullulan: industrially significant alpha-D-glucans. *Abstracts of Papers*, American Chemical Society, 172, Carb 70.

Jokay, L., G.E. Nelson & E.L. Powell (1967) Development of edible amylaceous coatings for foods. *Food Technol.*, **21**, 1064.

Jong Whan Rhim, Keum Taek Hwang, Hyun Jin Park & Soon Teck Jung (1996) Water vapor transfer characteristics of carrageenan-based edible films. *Korean J. Food Sci. Technol.*, **28**(3), 545–551 (in Korean).

Kader, A.A. (1986) Biochemical and physiological basis for effects of controlled and modified atmospheres on fruits and vegetables. *Food Technol.*, **40**, 99–104.

Kamper, S.L. & O. Fennema (1984) Water vapor permeability of an edible, fatty acid, bilayer film. *J. Food Sci.*, **49**(6), 1482–1485.

Kampf, N. & A. Nussinovitch (1998) Gum coating of cheeses. Osaka City University International Symposium Joint Meeting with the 4th International Conference of Hydrocolloids. 4–10 October, Osaka, Japan.

Kampf, N. & A. Nussinovitch (2000) Hydrocolloid coatings of cheeses. *Food Hydrocolloids*, **14**, 531–537.

Kampf, N., C. Zohar & A. Nussinovitch (2000a) Alginate coating of *X. laevis* embryos. *Biotechnol. Prog.*, **14**, 497–505.

Kampf, N., C. Zohar & A. Nussinovitch (2000b) Hydrocolloid coating of *X. laevis* embryos. *Biotechnol. Prog.*, **16**, 480–487.

Kaplan, H.J. (1986) Washing, waxing and color adding. In: (Wardowski, W.F., S. Nagy & W. Grierson, Eds.), *Fresh Citrus Fruits*. Avi Publishing Co., Westport, CT, p. 379.

Karel, M. (1975) Protective packaging of foods. In: (Fennema, O., Ed.), *Principles of Food Science*. Marcel Dekker, NY, pp. 399–464.

Kester, J.J. & O.R. Fennema (1986) Edible films and coatings: a review. *Food Technol.*, **December**, 47–59.

Kester, J.J. & O.R. Fennema (1988) Edible films and coatings: A review. *Food Technol.*, **42**, 47–59.

Kester, J.J. & O. Fennema (1989) An edible film of lipids and cellulose ethers: barrier properties to moisture vapor transmission and structural evaluation. *J. Food Sci.*, **54**(6), 1383–1389.

Kittur, F.S., K.R. Kumar & R.N. Tharanathan (1998) Functional packaging properties of chitosan films. *Food Res. Technol.*, **206**(1), 44–47 (in German).

Klose, R.E. & M. Glicksman (1972) Gums. In: (Furia, T.E., Ed.), *CRC Handbook of Food Additives*. 2nd Edn. CRC Press, Cranwood Parkway, Cleveland, OH, pp. 295–361.

Krochta, J.M., E.A. Baldwin & M. Nisperos-Carriedo (1994) *Edible Coatings and Films to Improve Food Quality*. Tachnomic Publishing Co. Inc., Lancaster, Basel.

Landrock, A.H. & B.E. Proctor (1952) Gas permeability of films. *Mod. Packag.*, **25**(10), 131–135, 199–201.

Lawrence, J.F. & J.R. Iyengar (1983) Determination of paraffin wax and mineral oil on fresh fruits and vegetables by high temperature gas chromatography. *J. Food Safety*, **5**, 119–124.

Lawson, J.A. (1960) Banana Packing and Waxing". *West Austr. Dept. Agr. Jour.* (Ser. 4), **1**, 41–45.

Lawton, J.W. (1996) Effect of starch type on the properties of starch-containing films. *Carbohydrate Polymers*, **29** (3), 203–208.

Li, T.X., Q.C. Zeng & N.H. He (1996). Film-forming properties of edible sodium alginate. *Sci. Technol. Food Ind.*, **6**, 4–8 (in Chinese).

Lowings, P.H. & D.G. Cutts (1982) The preservation of fresh fruits and vegetables. *Proc. Inst. Food Sci. Tech. Ann. Symp.*, July 1981, Nottingham, UK, p. 52.

Lui, T.X., Q.X. Zeng & H.H. He (1966) Film-forming properties of edible sodium alginate and application of the film. *Sci. Technol. Food Ind.*, **4**, 4–9 (in Chinese).

Magness, J.R. & H.C. Diehl (1924) Physiological studies on apples in storage. *J. Agric. Res.*, **27**, 1–38.

Martin Polo, M., C. Mauguin & A. Voilley (1992) Hydrophobic films and their efficiency against moisture transfer. Influence of the film preparation technique. *J. Agric. Food Chem.*, **40**(3), 407–412.

Mason, D.F. (1969) Fruit preservation. United States Patent 3,472,662.

Mathur, P.B. & H.C. Srivastava (1955) Effect of skin coatings on the storage behavior of mangoes. *Food Res.*, **20**, 559–566.

McCormick, R.D. (1975) Edible coating isolates oxygen and moisture, controls structure, seals in flavor. *Food Product Development*, **9**(4), 14, 16.

McHugh, T.H. (1996) Effects of macromolecular interactions on the permeability of composite edible films. In: *Macromolecular Interactions in Food Technology*. American Chemical Society, Washington DC, pp. 134–144.

McQuilken, M.P., J.M. Whipps & R.C. Cooke (1992) Use of oospore formulations of Pythium oligandrum for biological control of Pythium damping-off in cress. *J. Phytopathol.*, **135**(2), 125–134.

Mellenthin, W.M., P.M. Chen & D.M. Borgic (1982) In line application of porous wax coating materials to reduce friction discoloration of *Bartlett* and *d'Anjou* pears. *Hort. Sci.* **17**, 215–217.

Meyer, R.C., A.R. Winter & H.H. Wiser (1959) Edible protective coatings for extending the shelf life of poultry. *Food Technol.*, **13**, 146–148.

Miller, W.R., D. Chun, L.A. Risse, T.T. Hatton & T. Hinsch (1988) Influence of selected fungicide treatments to control the development of decay in waxed or film wrapped Florida grapefruit. *Proc. 6th Int. Cong.*, **3**, 1471–1477.

Mittal, K.L. (1983) *Adhesion Aspects of Polymeric Coatings*. Plenum Press, New York & London.

Moore, C.O. & J.W. Robinson (1968) Method for coating fruits. United States Patent 3,368,909.

Myllarinen, P., P. Rantamaki, J. Latva Koivisto & R. Ahvenainen (1997) Minimization of food packaging using active edible coatings. Possibilities and challenges. *VTT Research Notes*, no. 1840 (in Finnish).

Nakajima, K., T. Ikehara & T. Nishi (1996) Observation of gellan gum by scanning tunneling microscopy. *Carbohydrate Polymers*, **30**(2/3), 77–81.

Natrajan, N. & B.W. Sheldon (1995) Evaluation of bacteriocin-based packaging and edible film delivery systems to reduce Salmonella in fresh poultry. *Poultry Sci.*, **74**(Suppl. 1), 31.

Nelson, K.L. & O.R. Fennema (1991) MC films to prevent lipid migration in confectionery products. *J. Food Sci.*, **56**(2), 504–509.

Nelson, R.C. (1942) A method of studying the movement of respiratory gases through waxy coatings. *Plant Physiol.*, **17**, 509–514.

Niederhauser, D.O. & F.E. Bartell (1950) Report of Progress-Fundamental Research on the Occurrence and Recovery of Petroleum. Publishing of the American Petroleum Institute, The Lord Baltimore Press, Baltimore, p. 114.

Nisperos-Carriedo, M.O. & E.A. Baldwin (1988) Effect of two types of edible films on tomato fruit ripening. *Proc. Fla. State Hort. Soc.*, **101**, 217–220.

Nisperos-Carriedo, M.O, P.E. Shaw & E.A. Baldwin (1990) Changes in volatile flavor components of pineapple orange juice as influenced by the application of lipid and composite film. *J. Agric. Food Chem.*, **38**, 1382–1387.

Nussinovitch, A. (1994) Edible coatings of vegetables. *International Workshop on Gellan and Related Polysaccharides*. 14–15 November, Center for Cultural Exchange, Osaka City University, Osaka, Japan.

Nussinovitch, A. (1997) Agricultural uses of hydrocolloids. In: *Hydrocolloid Applications, Gum Technology in the Food and Other Industries*. Blackie Academic & Professional, London, UK, pp. 169–184.

Nussinovitch, A. (1998) Hydrocolloid coating of foods: a review. *The Leatherhead Food Industry J.*, **1**(3), 174–188.

Nussinovitch, A. (2000) Gums for coatings and adhesives. In: (Phillips, G. & P. Williams, Eds.), *Handbook of Hydrocolloids*. Woodhead Publishing Limited, London, UK.

Nussinovitch, A. & V. Hershko (1996) Gellan and alginate vegetable coatings. *Carbohydrate Polymers*, **30**, 185–192.

Nussinovitch, A. & N. Kampf (1992) Shelf-life extension and texture of alginate-coated mushrooms. IFTEC, 15–18 November, The Hague, The Netherlands.

Nussinovitch, A. & N. Kampf (1993) Shelf-life extension and conserved texture of alginate-coated mushrooms (*Agaricus bisporus*). *Lebensm.-Wiss. u.-Technol.*, **26**, 469–475.

Nussinovitch, A. & N. Kampf (2001) Hydrocolloid coating of cells. Patent Application No. 09/856,423.

Nussinovitch, A. & S. Lurie (1995) Edible coatings for fruits and vegetables. *Postharvest News and Information*, **6**(4), 53–57.

Nussinovitch, A., V. Hershko & H.D. Rabinowitch (2000) Protective coatings for food and agricultural products. Israel Patent 111495. United States Patent 6,068,867 and 6,299,915 (2001).

Paredes-Lopez, O., E. Camargo-Rubio & Y. Gallardo-Navarro (1974) Use of coatings of candelilla wax for the preservation of limes. *J. Sci. Food Agric.*, **25**, 1207–1210.

Park, H.J., K.H. Jo, J.W. Rhim & S.T. Jung (1996a) Salts and plasticizers effect on water vapor permeability of carrageenan-based biopolymer film. *IFT Annual Meeting: Book of Abstracts*, p. 127.

Park, J.W., R.F. Testin, H.J. Park, P.J. Vergano & C.L. Weller (1994) Fatty acid concentration effect on tensile strength, elongation and water vapor permeability of laminated edible films. *J. Food Sci.*, **59**(4), 916–919.

Park, J.W., R.F. Testin, P.J. Vergano, H.J. Park & C.L. Weller (1996b) Fatty acid distribution and its effect on oxygen permeability in laminated edible films. *J. Food Sci.*, **61**(2), 401–406.

Parris, N., D.R. Coffin, R.F. Joubran & H. Pessen (1995) Composition factors affecting the water vapor permeability and tensile properties of hydrophilic films. *J. Agric. Food Chem.*, **43**(6), 1432–1435.

Pearce, J.A. & C.G. Lavers (1949) Frozen storage of poultry. V. Effects of some processing factors on quality. *Canadian J. Research*, **27**, 253–265.

Pegg, R.B. & F. Shahidi (1999) Encapsulation and controlled release in food preservation. In: (Shafiur Rahman, M., Ed.), *Handbook of Food Preservation*. Marcel Dekker, Inc., New York and Basel, pp. 611–667.

Pennisi (1992) Sealed in edible film. *Sci. News*, **141**, p. 12.

Powell, G.M. (1980) Polyethylene glycole. In: (Davidson, R.L., Ed.), *Handbook of Water-Soluble Gums, and Resins*. McGraw-Hill Book Company, NY, pp. 18-1–18-31.

Psomiadou, E., I. Arvanitoyannis & N. Yamamoto (1996) Edible films made from natural resources, microcrystalline cellulose, methylcellulose and corn starch and polyols. *Carbohydrate Polymers*, **31**(4), 193–204.

Ptitchkina, N.M. & A. Ye. Chalykh (1994) Sorption of water vapour by K- and Ca-furcellaran films. *Food Hydrocolloids*, **8**(3–4), 251–258.

Quast, D.G. & M. Karel (1972) Technique for determining oxygen concentrations within packages. *J. Food Sci.*, **37**, 490–491.

Radnia, P.M. & J.W. Eckert (1988) Evaluation of imazalil efficacy in relation to fungicide formulation and wax formulation. *Proc. 6th Int. Cit. Cong.*, **3**, 1427–1434.

Ramon, A.A.T.V., M.S. Chinnan, M.C. Erickson & V.M. Balasubramaniam (1996) Lipid oxidation in roasted peanut coated with edible films. *IFT Annual Meeting: Book of Abstracts*, p. 18.

Rhim, J.W., H.J. Park, K.T. Hwang, S.G. Kang & S.T. Jung (1996) Lipid penetration of carrageenan-based films. *IFT Annual Meeting Book of Abstracts*, p. 127.

Rice, J. (1991) Biodegradable plastics, Do they have a future for food packaging applications? *Food Processing*, **52**(11), 34–35, 40, 42.

Rice, J. (1994) What's new in edible films? *Food Processing*, **55**(7), 61–62.

Rico-Pena, D.C. & J.A. Torres (1990a) Edible MC-based films as moisture-impermeable barriers in sundae ice cream cones. *J. Food Sci.*, **55**(5), 1468–1469.

Rico-Pena, D.C. & J.A. Torres (1991) Sorbic acid and potassium sorbate permeability of an edible MC-palmitic acid film: water activity and pH effects. *J. Food Sci.*, **56**(2), 497–499.

Rong Huei Chen, Jeun Hwang Lin & Mei Huaw Yang (1994) Relationships between the chain flexibilities of chitosan molecules and the physical properties of their casted films. *Carbohydrate Polymers*, **24**(1), 41–46.

Sapru, V. & T.P. Labuza (1994) Dispersed phase concentration effect on water vapor permeability in composite methyl cellulose-stearic acid edible films. *J. Food Process. Preserv.*, **18**(5), 359–368.

Shih, F.F. (1994) Interaction of soy isolate with polysaccharide and its effect on film properties. *Journal of the American Oil Chemists' Society*, **71**(11), 1281–1285.

Shih, F.F. (1996) Edible films from rice protein concentrate and pullulan. *Cereal Chem.*, **73**(3), 406–409.

Sinclair, W.B. (1961) *The Orange: Its Biochemistry and Physiology*. Berkeley, CA.

Smith, S.M. & J.R. Stow (1984) The potential of a sucrose ester coating material for improving the storage and shelf life qualities of Cox's Orange Pippin apples. *Ann. Appl. Biol.*, **104**, 383–391.

Smith, S., J. Geeson, K.M. Browne, P. Genge & H. Everson (1987) Modified atmosphere retail packaging of discovery apples. *J. Sci. Food Agric.*, **40**, 165–178.

Stelzer, G.I. & E.D. Klug (1980) Carboxymethylcellulose. In: (Davidson, R.L., Ed.), *Handbook of Water-Soluble Gums and Resins*. McGraw-Hill, NY, pp. 4.1–4.28.

Stern, S.A., T.P. Sinclair & T.P. Gareis (1964) An improved permeability apparatus of the variable-volume type. *Mod. Plastics*, **10**, 50–53.

Stoloff, L.S., J.F. Puncochar & H.E. Crowther (1948) Curb mackerel fillet rancidity. *Food Industries*, **20**, 1130–1132, 1258.

Sutherland, J.P., A.H. Varnam & M.G. Evans (1986) *A Color Atlas of Food Quality Control*. A Wolf Science Book, Weert, The Netherlands.

Swenson, H.A., J.C. Miers, T.H. Schultz & H.S. Owens (1953) Pectinate and pectate coatings. II. Application to nuts and fruit products. *Food Technol.*, **7**, 232.

Tanaka, T. (1981) Gels. *Sci. Am.*, **245**(1), 110–123.

Torres, J.A., M. Motoki & M. Karel (1985) Microbial stabilization of intermediate moisture food surfaces. Control of surface preservative concentration. *J. Food Process Preserv.*, **9**, 75.

Toulmin, H.A. Jr. (1956a) Method of preserving shrimp. United States Patent 2,758,929.

Toulmin, H.A. Jr. (1956b) Method of preserving shrimp. United States Patent 2,758,930.

Toulmin, H.A. Jr. (1957) Method of preserving food products. United States Patent 2,790,721.

Trout, S.A., E.G. Hall & S.M. Sykes (1953) Effects of skin coatings on the behavior of apples in storage. *Aust. J. Agr. Res.*, **4**, 57–81.

Vernon Carter, E.J. & P. Sherman (1981) Rheological properties and applications of mesquite tree (*Prosopis juliflora*) gum. IV. Rheological properties of mesquite gum films at the oil-water interface. *J. Dispersion Sci. Technol.*, **2**(4), 399–413.

Warburton, S.C., A.M. Donald & A.C. Smith (1993) The deformation of thin films made from extruded starch. *Carbohydrate Polymers*, **21**(1), 17–21.

Ward, G. & A. Nussinovitch (1996) Gloss properties and surface and surface morphology relationship of fruits. *J. Food Sci.*, **61**(5), 973–977.

Ward, G. & A. Nussinovitch (1997) Characterizing the gloss properties of hydrocolloid films. *Food Hydrocolloids*, **11**(4), 357–365.

Warth, A.H. (1986) *The Chemistry and Technology of Waxes*. Reinhold Pub. Co., New York, pp. 37–192.

Watson, E.L. & J.C. Harper (1983) *Elements of Food Engineering*. Van Nostrand Reinhold, NY, pp. 252–285.

Wicks, Z.W., Jones, N. Frank & S. Peter-Pappas (1994) *Organic Coatings, Science and Technology*. John Wiley and Sons, Inc., New York.

Williams, S.K., J.L. Oblinger & R.L. West (1978) Evaluation of a calcium alginate film for use on beef cuts. *J. Food Sci.*, **43**, 292–296.

Willis, R.H., T.H. Lee, D. Graham, W.B. McGlasson & E.G. Hall (1981) *Postharvest, an Introduction to the Physiology and Handling of Fruit and Vegetables*. Avi Publishing Co. Inc., Westport, CN, pp. 1–2.

Wong, D.W.S., F.A. Gastineau, K.S. Gregorski, S.J. Tillin & A.E. Pavlath (1992) Chitosan lipid films: microstructure and surface energy. *J. Agric. Food Chem.*, **40**(4), 540–544.

Wong, D.W.S., K.S. Gregorski, J.S. Hudson & A.E. Pavlath (1996) Calcium alginate films: thermal properties and permeability to sorbate and ascorbate. *J. Food Sci.*, **61**(2), 337–341.

Wu, S. (1973) Polar and nonpolar interactions in adhesion. *J. Adhesion*, **5**, 39–55.

Wu, M.T. & D.K. Salunkhe (1972) Control of chlorophyll and solanine synthesis and sprouting of potato tubers by hot paraffin wax. *J. Food Sci.*, **37**, 629–630.

Yuen, S. (1974) Pullulan and its applications. *Process Biochem.*, **9**(9), 7–9, 22.

Zhuang, R., L.R. Beuchat, M.S. Chinnan, R.L. Shewfelt & Y.W. Huang (1996) Inactivation of *Salmonella montevideo* on tomatoes by applying cellulose-based edible films. *J. Food Protection*, **59**(8), 808–812.

Zoebel, E.H. (1991) Edible packaging from starch. *Kaka-und-Zucker*, **43**(5), 4–7 (in German).

Chapter 3
Dry Macro- and Liquid-Core Hydrocolloid Capsules

3.1 Introduction

Encapsulation, micro or macro, is a specialized form of edible packaging. In food processes utilizing encapsulation, the approach is to apply the encapsulation process only to those ingredients which are unstable, volatile or particularly reactive. Thus the provision of an envelope around these ingredients provides stability and protection for the whole product (Daniels, 1973).

3.2 Soft gelatin capsules

Liquid foods, as well as instant (soluble) coffee and other food powders, can be conveniently contained in a gelatin capsule (Maddox, 1971). The interior of the capsule contains a suitable instant food which dissolves or disperses promptly upon addition of water (Fig. 3.1). The capsule is maintained in a dry form in a suitable enclosure, such as a hermetically sealed bottle, blister-pack packaging or the like, until use. Soft gelatin capsules are commonly used in pharmaceuticals, cosmetics and food supplements (Moorhouse & Grundon, 1994). Gelatin is the basic capsule shell component and it is formulated with suitable ingredients to encapsulate a wide variety of materials. Gelatin's special properties are of particular interest in foods since the material acts as a barrier and protects liquid capsule contents from the outside environment. On the one hand, it acts as a physical barrier to bacteria, yeasts and molds. On the other, it provides a low-permeability membrane to gases. The gelatin shell is transparent, can be formed in a wide range of sizes and shapes and dissolves quickly in hot water, releasing its encapsulated liquid (Moorhouse & Grundon, 1994). In the process of encapsulation, two continuous strips of molten gelatin are brought together simultaneously and the fill formulation is injected between them. Size and shape can be determined by the die roll. The advantages of encapsulation are: portion control, easy use and storage, extended shelf-life, improved aesthetic appeal, the variety of sizes available, disposability and edibility, improved product aromatics versus time, and biodegradability. A few food applications are: *real chicken broth* capsules which retain and deliver flavor more effectively than the powder system, encapsulated lemon oil for meringue pie mix, mint essence

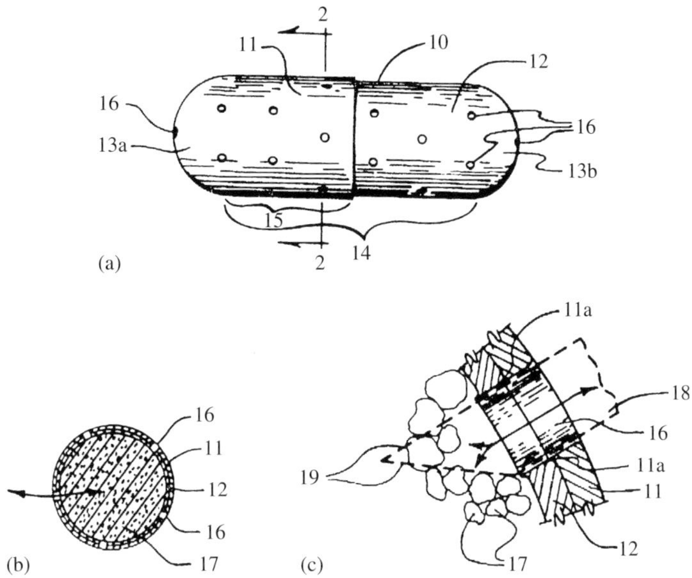

Fig. 3.1 Gelatin capsule: (a) A capsule having a cylindrical cap part and a matching telescopically engaged body. (b) The perforations provide open communication from the exterior of the capsule to the interior. (c) As water comes into contact with the capsule it enters the several perforations, passes through the capsule walls and comes into contact with the food content. Since instant foods are more soluble or dispersible in water than is gelatin, the usual result is that the food contents begin to dissolve or disperse first rather than the gelatin. (Adapted from Maddox, 1971.)

capsules for the tinned goods market (Moorhouse & Grundon, 1994). Future possibilities could be inclusion of the capsules as part of dry mixes, instant foods and microwaveable products.

Machines have been invented to cost-effectively produce seamless gelatin capsules, to achieve excellent shell clarity and to deliver fills to an exacting standard of accuracy. A wide range of filler materials can be encapsulated within these capsules, such as most vegetable oils, essential oils and fish oils, as well as suspensions of crystalline materials milled with oils. Materials include: fish oils, multivitamin suspensions, halibut liver oil, mouthwashes, oil of evening primrose, antitussive preparations, wheat germ oil, flavor oils, vitamin E, unsaturated fatty acids, vitamin A and D, insecticides, vitamin K, perfume oils, inhalation and garlic oils (ITS Machinery Development Ltd., 1996).

3.3 Liquid-core capsules

Liquid-core hydrocolloid capsules are liquids encapsulated in a spherical polymer membrane (Vergnaud, 1992). Lim and Sun (1980) were the first to describe a method of producing alginate–polylysine liquid-core micro-capsules to encapsulate pancreatic islets. Production of these capsules included suspending cells in a sodium alginate solution, forming small spherical calcium alginate beads by cross-linking with calcium salt, and reacting with polylysine to create a polylysine alginate membrane around the bead. In the final stage, the bead's core, composed of calcium alginate gel, was solubilized, thus forming a liquid-core micro-capsule containing cells (Lim & Sun, 1980). With this procedure, cells could also be found in the membrane matrix, leading to the proposal of an approach to eliminate this possibility (Wong & Chang, 1991). In the latter approach, cells were entrapped in alginate-gel micro-spheres, which in turn were contained within larger beads, resulting in a greater distance between the cells and the surface of the larger alginate bead. Similar to Lim and Sun's (1980) procedure, the surface of the larger micro-sphere was reacted with poly-L-lysine and then with alginate to form a coating membrane. The contents of the micro-capsule were then liquefied with sodium citrate to remove the calcium from the array. The cells in the smaller entrapped gel micro-sphere were released and allowed to float freely in the liquid core of the resultant beads (Wong & Chang, 1991). In both cases, the production of liquid-core hydrocolloid capsules consisted of several stages. A one-step technique to form and modify the mechanical and slow-release properties of liquid-core hydrocolloid capsules (Fig. 3.2) was then described (Nussinovitch, 1992, 1994; Nussinovitch *et al.*, 1996). Proteins with molecular masses of 2500–205 000 Da were entrapped within liquid-core alginate, alginate–chitosan or alginate–polylysine capsules. The ratio, $R(t)$, between protein concentration of a certain protein in the external fluid into which capsules were immersed, at a given time (t), and the evaluated equilibrium concentration of that same protein in the external fluid, was

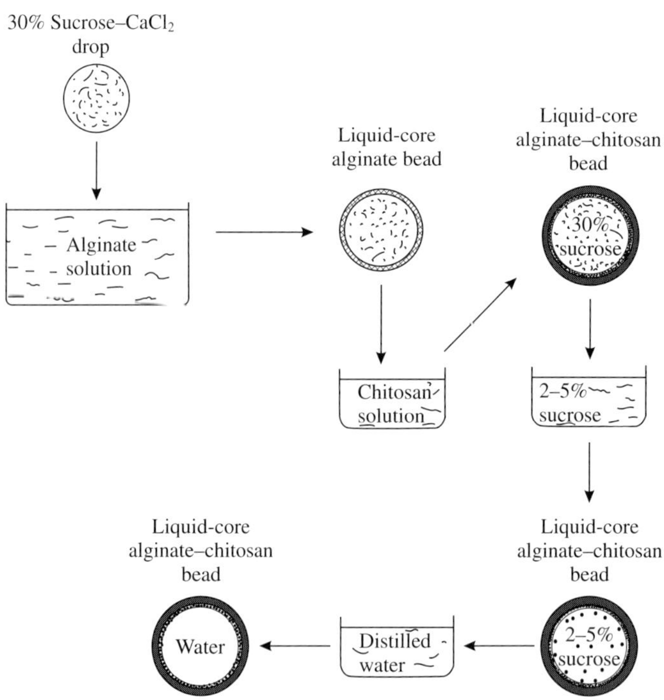

Fig. 3.2 A method to produce various liquid-core capsules. (From Nussinovitch *et al.*, 1996, with permission from Elsevier Science.)

calculated and plotted against time. These diffusion results were compared to diffusion from *whole* alginate beads. Better slow-release properties were achieved when polylysine or chitosan was used to change the permeability of the alginate membrane. These chemical treatments also strengthened the membrane and affected its brittleness (Nussinovitch *et al.*, 1996).

Liquid-core spherical alginate–chitosan capsules, with various hydrocolloid concentrations within their membrane, were produced in a single step (Fig. 3.2). This simplified process is advantageous over other multi-stage methods of fluid-core capsule production. The contents of the capsule were either distilled water or sucrose solutions (2.5 and 30%, w/w), although other viscous liquids can be used. Beads with 0, 2 and 5% sucrose were produced by diffusion of sucrose out of liquid-core capsules containing 30% sucrose. The spherical shape of the capsule was retained after diffusion. Mechanical properties of various capsules were studied after incubation at 25, 37, 45, 55 and 85°C for 5, 30 and 60 min, while those of the liquid-core water capsules were studied for a further 2 weeks at 25°C. Capsules with a higher hydrocolloid concentration within their membrane

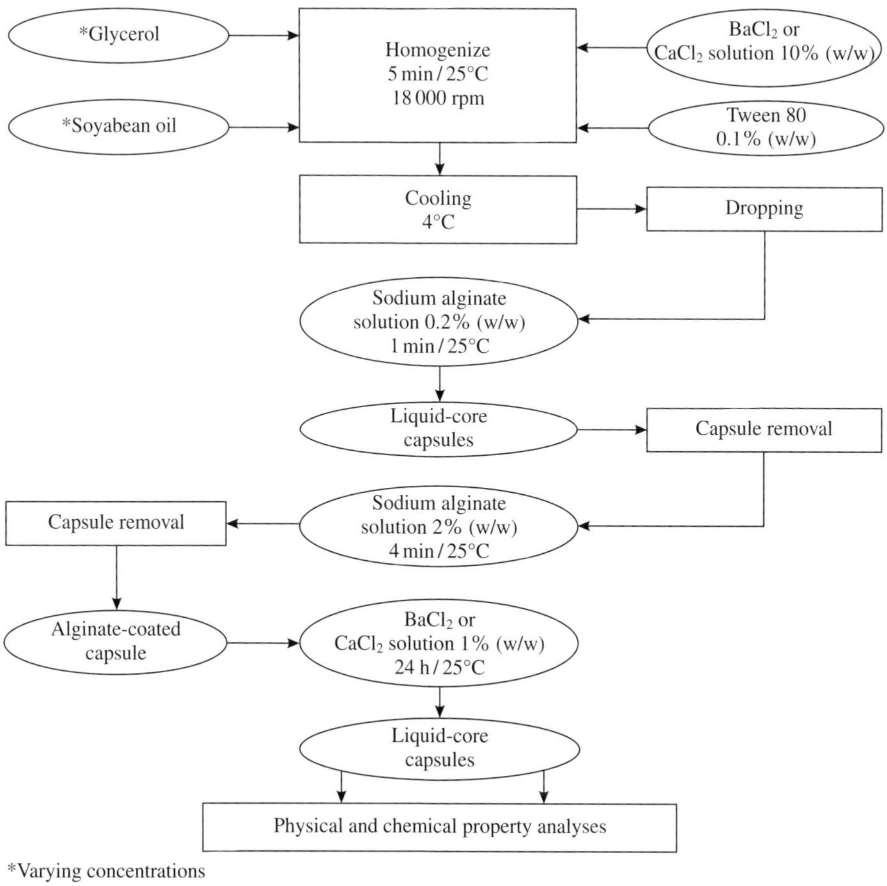

Fig. 3.3 Production of liquid-core hydrocolloid oil capsules. (From Nussinovitch & Solomon, 1998.)

displayed more stress at failure (strength) and less brittleness than those with lesser solid membrane content. Following diffusion, capsules with 2 and 5% sucrose were weak compared to those with 30% sucrose; however, no membrane rupture was observed after incubation. The weakest capsules were those containing water incubated for long times at higher temperatures. The temperature–stability relationship of the liquid-core capsules was presented. Projection of three-dimensional curves of mechanical property versus time, temperature or percent sugar, offers a convenient way of examining the desired mechanical properties and their dependence on liquid-core composition and incubation conditions (Nussinovitch *et al.*, 1997).

3.4 Liquid-core hydrocolloid oil capsules

A one-step method to produce liquid-core hydrocolloid oil capsules, 4–5 mm in diameter with a 70–90 μm thick membrane, was described (Nussinovitch & Solomon, 1998). Three basic types of beads were produced, containing 40–90% glycerol, 0 to 50% soybean oil and 10% of either calcium chloride or barium chloride (Fig. 3.3). Membranes were first strengthened after capsule formation with Ca or Ba cations, followed by immersion in 2% sodium alginate and strengthening of this alginate layer with either calcium chloride or barium chloride. The higher the oil content within the liquid composing the capsule, the lower its liquid medium's density. Although the aspect ratio for the oil-core hydrocolloid capsules ranged from 0.91 to 0.97, most capsules had a value of ~0.95, reflecting their high degree of sphericity. The smallest liquid-core volume was recorded for the capsules with the least included oil, and for those which underwent further cross-linking and contained

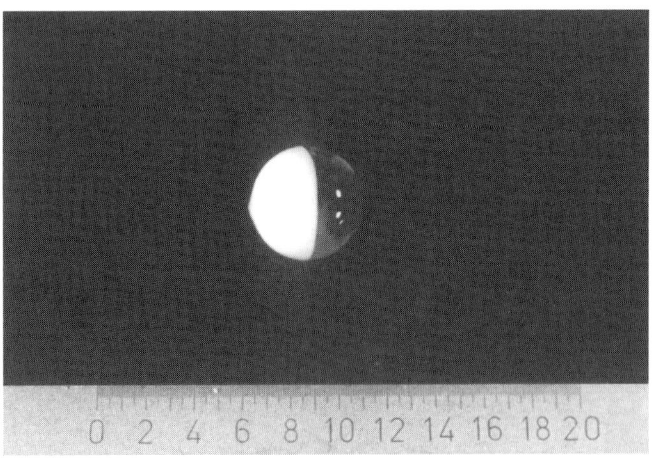

Fig. 3.4 Liquid-core hydrocolloid oil capsule consisting of glycerol and 50% soyabean oil before capsule shrinkage. The *white phase* is the oil. This capsule has a 20.7-mm^2 maximum projection area, an average volume of 71.6 mm^3, an equivalent diameter of 5.1 mm and an aspect ratio of 0.96 reflecting its high degree of sphericity. (From Nussinovitch & Solomon, 1998.)

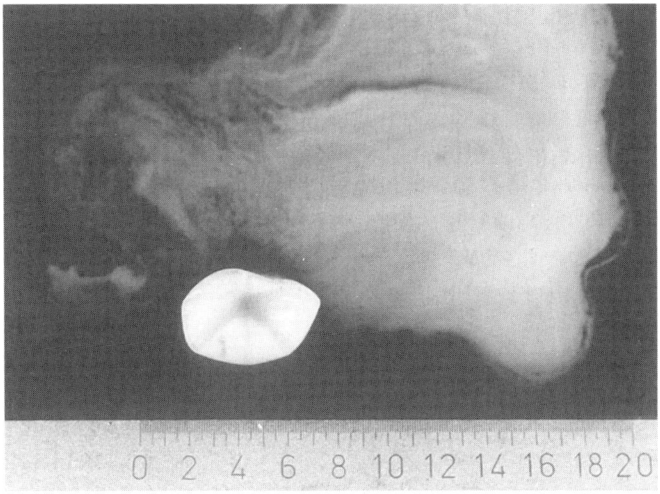

Fig. 3.5 Liquid-core hydrocolloid oil capsule after compression and bursting. The *pseudo-stress* and engineering strain at failure of a 50% hydrocolloid oil capsule was 40 kPa and 0.7, respectively. (From Nussinovitch & Solomon, 1998.)

25% included oil (Nussinovitch & Solomon, 1998). The maximum projected area was smallest with the lowest oil content and Ba cross-linking, and was a function of time. The longer the time, the smaller the capsule. In general, the higher the oil content within the capsule, the weaker the capsule. Membrane strengthened with Ba^{2+} was always stronger than that strengthened with Ca^{2+}. With Ba^{2+} capsules, respective stresses at failure were 185, 97 and 64 kPa for 0, 25 and 50% included oil. With Ca^{2+} oil-core beads, strengths were 136, 66 and 40 kPa for 0, 25 and 50% included oil, respectively (Fig. 3.4). None of the capsules were brittle, their engineering strain at failure deviating between 0.79 and 0.78. The deformability modulus calculated at a strain value of about 0.12 (when the highest R^2 of σ versus ε was achieved) followed the same trend as strength. In other words, capsules were stiffer when membranes were strengthened with Ba^{2+} versus Ca^{2+}. Moreover, the higher the oil content within the capsule, the lesser the stiffness of the membrane and capsule (Fig. 3.5) (Nussinovitch & Solomon, 1998). Since the oil is entrapped within the capsule, it is assumed that capsules with even higher oil content can be achieved by changing the cross-linking solution or using other liquids that could further shrink the capsule, such as acetone. Liquid-core hydrocolloid oil capsules can be used for food and non-food purposes, when oil-soluble ingredients are included.

In other studies on micro-encapsulation by coagulation, soybean oil was used as the core material and gum arabic and gelatin as shell materials. Conditions for encapsulation were optimized at 1% gum arabic and gelatin, pH 4.4, 40°C, and a mixing time of 15 min. Under these conditions, inclusion rate was greater than 50%. Microwave de-watering was achieved in 5–6 min in a 500 W microwave oven for a 2 cm thick layer of micro-capsules (Wang, 1997).

3.5 Biotechnological applications of liquid-core capsules

Descriptions of food or food-related biotechnological applications for liquid-core capsules in the literature are scarce. They focus mainly on cell inclusion within liquid-core capsules for specific material production. Immobilization of *Lactobacillus casei* subsp. *rhamnosus* ATCC 10863 in calcium alginate beads, and lactic acid production by the immobilized cells are described (Yoo *et al.*, 1996). The alginate beads (capsules) contain an interphasic membrane and a liquid core; these beads allow more space for cell growth than solid gel beads and can contain 1.5-fold higher cell numbers. Barium alginate gel beads were more stable than calcium alginate beads, and bead stability was further enhanced by treatment with chitosan and barium chloride. Stable lactic acid productivities of greater than $2.7\,gl^{-1}h^{-1}$ were achieved during fermentation of *L. casei* cells immobilized in chitosan-coated barium alginate capsules. Cell leakage from the capsules was relatively low during repeat batch fermentations (Yoo *et al.*, 1996).

Liquid-core calcium alginate capsules have also been used to immobilize recombinant yeast cells with a plasmid containing the SUC2 gene encoding invertase. Invertase activity in immobilized cells was slightly higher than that in free cells contained within a batch system. Temperature for optimal enzyme activity was 65°C, in both entrapped and free cultures. However, the immobilized preparation should improve thermal activity. Immobilized cells retained more than 90% activity for 7 days at 30°C. Both a stirred-tank reactor and a packed-bed reactor are options for continuous sucrose hydrolysis (Chang *et al.*, 1996).

3.6 Special food applications

3.6.1 Jelly-like foods

Natural gums are used in the confectionery industry. At one time, agar was used for the production of jellies (candies) and marshmallows, and gum arabic was used in gumdrops. The gum within the formulation served to form the *jelly*, but an additional function was to prevent sugar crystallization and to emulsify fat, keeping it evenly distributed within the product (Klose & Glicksman, 1975). A special ice confection consisting of jelly balls with an elastic skin and liquid core is a specific product that contains polysaccharides obtained from *Alcaligenes* or *Agrobacterium* cultures grown on glucose, with sucrose, milk and vanilla flavoring. These liquid-core elastic skin jelly balls are then frozen to form an ice confection (Kimura *et al.*, 1975). These ice confections can include fruit jellies or milk jellies (Takeda Yakuhin Kogyo, 1974). As stated, these jelly-like products are prepared from a polysaccharide produced by Takeda Chemical Industries Ltd. (Tokyo, Japan). The gum powder swells and gels when added to water and heated. Its gels are thermally irreversible and unaffected by further addition of water and can be produced over a pH range of 2.0–9.5 in the presence of many food additives. The gels may be used to make novel food products consisting of a jelly-like skin with a liquid core, and canned jellies. The concentration of the polysaccharide in water must be greater than 1.5% for gel stability and less than 0.6% for taste acceptability. The gels are freeze–thaw stable and may be used to make an ice confection contained in an elastic gel skin (Anon, 1977).

3.6.2 Fruit products

A combination of compression and shearing forces is used to extract juice from fruits or vegetables. For pulp production, and in the case of grapes, tomatoes or other soft fruits, are heated, if necessary, to soften their tissues and pulp is forced through the perforations of the pulping equipment's screen, the size of which determines the consistency of the resultant product (Fellows, 2000). Unique uses of such fruit products (i.e. juice, pulp or puree) for production of soft viscous, fruit-based, membrane-coated items by a membrane were described decades ago. For example fruit pulp, puree or juice containing soluble Ca salt is extruded to form drops which are coated with a thin skin of alginate or pectate sol. The coated drops are exposed to an aqueous setting bath containing a soluble Ca salt (Sneath, 1975). Drops of aqueous fruit material are coated with an aqueous alginate or pectate solution and dipped in a solution containing Ca or Al ions to gel the surface. Another report deals with food products which simulate the non-uniform structure (soft core and tough skin) of soft fruit, e.g. black and red currants; these are obtained by surrounding droplets of, for example, fruit pulp or juice containing Ca or Al ions with an alginate or pectinate sol, and treating the coated droplets in a gelatinizing bath containing Ca or Al ions. Coated droplets are obtained by co-extrusion, the droplets being dipped directly into the coagulating bath (Unilever, 1974, 1980).

3.6.3 Encapsulating aroma and health compounds

Liquid-core hydrocolloid capsules can include coffee aroma, β-carotene, or any liquid that does not readily dissolve the shell. A procedure for encapsulating volatile aroma compounds is described. The method involves dissolving an inert gas (such as CO_2) in an aromatized edible liquid, then co-extruding the gas/liquid mixture with a molten carbohydrate material which has a glass transition temperature of 20–80°C. The result is a continuous stream of carbohydrate outer shell surrounding an inner core of aromatized, gasified liquid. This stream is extruded into a pressure chamber, at a pressure higher than that of the inert gas in the aromatized liquid core, until the carbohydrate shell cools to below the glass transition temperature and hardens. When these capsules, which can be incorporated directly into an instant consumer product, contact hot water, the inert gas in the inner core expands and ruptures the capsule wall, releasing the aromatized liquid. Coffee aroma capsules are prepared from 100% coffee-derived material, in which coffee aroma frost is dissolved in coffee oil and gasified, then encapsulated in an outer shell of hardened amorphous coffee glass (Garwood *et al.*, 1995). A health food containing encapsulated *Dunaliella* algae is manufactured by adding cyclodextrin (chemically and physically stable molecules formed by the enzymatic modification of starch), to *Dunaliella* powder, stirring and blending, adding an antioxidant (vitamin A or E), lubricant (talc, esters of sucrose with a fatty acid) and a binder (a sugar), blending, granulating, and encapsulating in a light-impermeable hard capsule. A variant employing a soft capsule is also described. The food is rich in β-carotene (Tanaka, 1990). Micro-capsules with a solid, fusible shell and multiple liquid cores (of any liquid, e.g. an aqueous solution, which does not readily dissolve the shell) can be produced by spray-cooling a water-in-oil emulsion. The resulting dry, free-flowing powder can be heated or otherwise processed to release the contents of the micro-capsules (Morgan & Blagdon, 1993).

3.6.4 Other foods

Reports or patents discussing specialized or common foods that cannot be regarded under one unifying title but have liquid-core capsules in common, can be located in the literature. A few examples of such products that are not directly related are listed briefly in this section. One report describes a process for preparing encapsulated foods and drinks filled with a desired edible liquid (Ueda, 1985). A core liquid is prepared by adding Ca salt and if necessary, other additives, to sugar liquid. Membranes of mainly calcium alginate are formed on the surface of the core liquid by dropping it into alginic acid salt liquid, thereby surrounding it with these membranes to form capsules. The core liquid inside the capsules is then exchanged with water or other digestible liquid by immersing the capsules in the appropriate liquid. Processes are described for the production of encapsulated dressing, fruit juice, alcoholic drinks, liquid sweetening materials and cut solid foods, e.g. vegetables (Ueda, 1985). Production of a roe-like, multi-layer, spherical structure composed of greater than two layers has been described. It involves discharging two edible materials through a multi-tubular nozzle to form two sol materials having different properties, with the material forming the inner layer being convertible to an aqueous gel that is softer than the outer material layer (Kuwabara & Jyouraku, 1983). Another patent describes novel ice cream products that contain small beads which are solid or hollow with a liquid center. These hollow capsules are based on either alginate or pectate with low-methoxy content (Unilever, 1979). Inclusion of liquid-core capsules within a food enables the introduction of a blend of textures and tastes in the same bite and is expected to develop in the future, when more sophisticated fun foods hit the market.

3.7 Dry hydrocolloid capsules

Dry hollow capsules were produced by freeze-drying liquid-core hydrocolloid capsules. Although related to their wet counterparts, these capsules differ in their basic characteristics as well as in their potential uses in food, medicine or other pharmaceutical products. Water is sublimated during freeze-dehydration, leaving a dry hollow hydrocolloid capsule with its inner membrane coated with sedimented guar gum. Both capsules maintained their integrity and spherical shape, although a small decrease in volume was observed relative to the wet capsules. The outer structure of the dry hollow capsules was rough, with many *folds* and *wrinkles*. Mini-cracks were

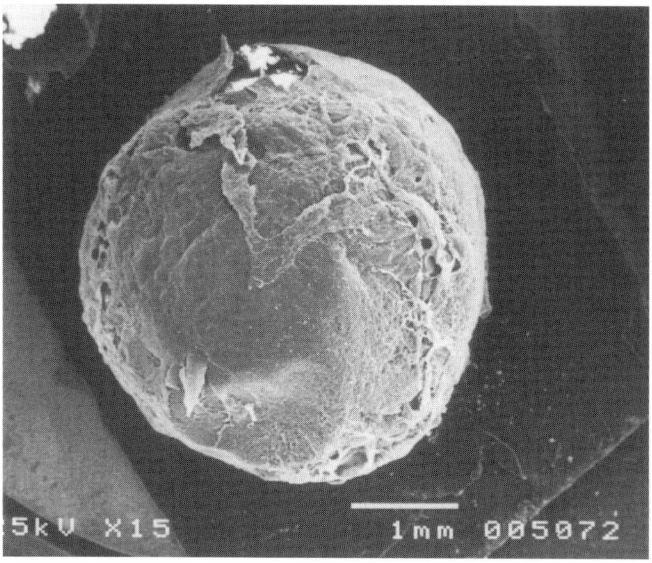

Fig. 3.6 SEM micrograph of a dry macro-capsule which can be used for encapsulating special food ingredients.

observed which nevertheless did not disturb the capsule's integrity (Fig. 3.6). Future studies into improving their strength and eliminating the mini-cracks on their surface could lead to their use in situations in which separate inner and outer surfaces are required: for example, different microorganisms could be immobilized or attached to separate surfaces at the same time (Nussinovitch *et al.*, 1999; Regev *et al.*, 1998).

By drying macro-capsules containing a starch suspension, one can achieve, in one step, a cellular spongy structure with a membranous coating, which can act as a system for controlled release. The physical properties of the capsules can be controlled by including different fillers such as talc, bentonite, silicon oxide, calcium carbonate or kaolin. The influence of the fillers on the liquid-core capsules before drying was to increase brittleness and decrease strength and hardness. After drying, the filler particles appeared to be entrapped within the starch matrix, while some of them covered it. Analysis of the relative amounts of the different elements (building the filler) within the capsule revealed that in the capsule containing 5% filler, their concentrations increased closer to the center. In those capsules containing 10% filler, the elements were distributed more or less equally across the internal capsule area. Stress–strain relationships of the compressed capsules resembled those of other cellular solids. No shoulder to those curves (see Chapter 8) was observed. After drying, the capsules containing the fillers were stronger than the controls. In most cases, the higher the filler concentration, the stronger the capsules. This may have been due to the higher concentration of solid constituents within the capsule structure. Sucrose embedded within a blank capsule diffused out within 1 h, in comparison to capsules that included fillers, where even after 6 h the sucrose had not been completely released (Nussinovitch & Regev, 1998; Regev *et al.*, 1998).

3.8 References

Anon (1977) Fascinating jelly-like foods. *Food Eng. Int.*, **2**(4), 38–39.

Chang, H.N., G.H. Seong, I.K. Yoo, J.K. Park & J.H. Seo (1996) Microencapsulation of recombinant *Saccharomyces cerevisiae* cells with invertase activity in liquid-core alginate capsules. *Biotechnol. Bioeng.*, **51**(2), 157–162.

Daniels, R. (1973) *Edible Coatings and Soluble Packaging*. Noyes Data Corporation, London.

Fellows, P. (2000) *Food Processing Technology Principles and Practice*. 2nd edition, CRC Press, Boca Raton, FL.

Garwood, R.E., Z.I. Mandralis & S.A. Westfall (1995) Encapsulation of volatile aroma compounds. United States Patent 5,399,368.

ITS Machinery Development Ltd. (1996) Advanced technology in seamless encapsulation. Globex Mark III Brochure, pp. 1–4.

Kimura, H., K. Kusakabe, S. Sato & H. Nakatani (1975) Jelly-like foods. United States Patent 3,908,027.

Klose, R. & M. Glicksman (1975) Gums. *Handbook of Food Additives*. 2nd edition, Academic Press, 295–358.

Kuwabara, K. & M. Jyouraku (1983) Process for production of roe-like multilayer spherical structure. United States Patent 4,375,481.

Lim, F. & A.M. Sun (1980) Microencapsulated islets as bioartificial endocrine pancreas. *Science*, **210**, 908–910.

Maddox, V.H. (1971) Gelatin capsule. United States Patent 3,620,759.

Moorhouse, S. & V. Grundon (1994) Encapsulating: A new concept for the food industry. *Nutr. Food Sci.*, **2**, 17–19.

Morgan, R. & P.A. Blagdon (1993) Methods of encapsulating liquids in fatty matrices, and products thereof. United States Patent 5,204,029.

Nussinovitch, A. (1992) Thermostable liquid cells. Israel Patent Application 103,354.

Nussinovitch, A. (1997) Temperature-stable liquid cells. United States Patent 6,099,876.

Nussinovitch, A. & G. Regev (1998) Cellular macro-capsules. Provisional Application 60/111,643, filed 09/12/1988.

Nussinovitch, A. & A. Solomon (1998) Liquid-core hydrocolloid-oil capsules. In: (Williams, P.A. & G.O. Phillips, Eds.), *Gums and Stabilizers for the Food Industry*. The Royal Society of Chemistry, Cambridge, **9**, 323–332.

Nussinovitch, A., Z. Gershon & M. Nussinovitch (1996) Liquid-core hydrocolloid capsules. *Food Hydrocolloids*, **10**(1), 21–26.

Nussinovitch, A., Z. Gershon & M. Nussinovitch (1997) Temperature-stable liquid-core hydrocolloid capsules. *Food Hydrocolloids*, **11**(2), 209–215.

Nussinovitch, A., G. Regev & S. Chen (1999) Cellular hydrocolloid macrocapsules. IFT Annual Meeting and Food Expo, July 24–28, Chicago, IL, USA.

Regev, G., Y. Ungar & A. Nussinovitch (1998) Cellular dry macro-capsules. *Bioencapsulation VII and Mini-Symposium on Microencapsulation*. Tidewater Inn, November 20–23. Easton, MD, USA.

Sneath, M.E. (1975) Simulated soft fruits. United States Patent 3,922,360.

Takeda Yakuhin Kogyo, K.K. (1974) Jelly-like foods. British Patent 1,360,465.

Tanaka, Y. (1990) Process for production of encapsulated foodstuff containing *Dunaliella* algae. United States Patent 4,915,965.

Ueda, T. (1985) Process for preparing edible products in the form of capsules. United States Patent 4,507,327.

Unilever Ltd. (1980) Encapsulated fruit product. British Patent 1,564,452.

Unilever, N.V. (1974) Method for encapsulating a liquid. Netherlands Patent Application 7,400,889 (in Dutch).

Unilever, N.V. (1979) Ice-cream product and method of manufacture. Netherlands Patent Application 7,808,870 (in Dutch).

Vergnaud, J.M. (1992) In: *Drying of Polymeric and Solid Materials*. Springer Verlag, London, Berlin, p. 127.

Wang, X.L. (1997) Study on microencapsulation by coagulating method. *J. Zhengzhou Grain Coll.*, **18**(3), 29–34 (in Chinese).

Wong, H. & T.M.S. Chang (1991) A novel two step procedure for immobilization living cells in microcapsules for improving xenograft survival. *J. Biomater. Artif. Cells Immobil. Biotechnol.*, **19**, 687–690.

Yoo, I.K., G.H. Seong, H.N. Chang & J.K. Park (1996) Encapsulation of *Lactobacillus casei* cells in liquid-core alginate capsules for lactic acid production. *Enzyme Microb. Technol.*, **19**(6), 428–433.

Chapter 4
Multi-Layered Hydrocolloid Products

4.1 Introduction

A simple way of achieving different textures and tastes in the same bite is to construct a food product made up of different layers. A few such multi-layered food products are already on the market, for example, crunchy wafers that include a sweet vegetable-fat-based chocolate or vanilla taste filling between the brittle wafers and a multi-layered, sweetened, agar-based confection for children. In the latter, the textures of the layers are similar, but their tastes and colors can differ. In the Orient, where the awareness of different gel textures is much more developed than in the West, a sweet curdlan-based multi-layered gel has been developed (Ikeda *et al.*, 1976). In that case, although the layers are built from the same hydrocolloid (curdlan), two different layer types are prepared from its powder by heating the suspension to different temperatures. Multi-layered hydrocolloid-based foods are important in the framework of foods of the future. Mathematical models predicting the mechanical properties of multi-layered gels, foods and other man-made products have already been proposed. One example of these is a model for predicting the stress–strain relationship of layered polymeric sponges made of polyurethane (Swyngedau *et al.*, 1991b). Another model (Nussinovitch *et al.*, 1991) was developed to calculate the compressive deformability of double-layered curdlan gels. We used the latter model to predict the compressive deformabilities of gels, the layers of which were glued together by one of the three adhesion techniques (Ben-Zion & Nussinovitch, 1995).

4.2 Deformability modulus and compressive deformabilities of multi-layered gels and texturized fruits

It is possible to predict the deformability modulus (stiffness) of a layered gel array, prepared with various gelling agents and other ingredients. The prediction, based on the deformability moduli and heights of the separate layers, is reasonably accurate and the methodology is generally applicable for predicting the stiffness of multi-layered gels, texturized fruits (Fig. 4.1) and other such food products (Ben-Zion & Nussinovitch, 1996). E_{D_A}, E_{D_B}, and $E_{D_{AB}}$ correspond to the deformability moduli of layers A, B and the whole array, respectively, in a combined gel consisting of two

Fig. 4.1 Four-layered texturized fruit based on agar, LBG and fruit pulp. (From Ben-Zion & Nussinovitch, 1996.)

layers. Ho_{0_A}, Ho_{0_B} and $Ho_{0_{AB}}$ are the heights of layers A, B and their addition, and the stress is defined as: $\sigma = E_D\epsilon$. These moduli can be calculated using Equations 1 and 2. Equation 1 is correct when engineering strain is being considered, whereas Equation 2 is applicable when Hencky's strain is being calculated.

$$E_{D_{AB}} = \frac{1}{(H_{0_A}/H_{0_{AB}}E_{D_A} + H_{0_B}/H_{0_{AB}}E_{D_B})} \quad (1)$$

$$E_{D_{AB}} = \frac{\sigma}{\ln\left[\frac{H_{0_A}}{H_{0_{AB}}}e^{\frac{\sigma}{E_{D_A}}} + \frac{H_{0_B}}{H_{0_{AB}}}e^{\frac{\sigma}{E_{D_B}}}\right]} \quad (2)$$

This empirical mathematical model successfully predicted the deformability modulus of a multi-layered gel composed of agar and one of four galactomannans in three different layer-thickness combinations, and of a four-layered gel array of texturized fruit (see Chapter 7). The model was based on the assumption that the uniaxial stress in the layers is the same and that their deformations are additive. No significant differences were found between experimental deformability moduli and those predicted by the model. Both Equations 1 and 2 gave more or less similar estimates of predicted E_D values and are simply expanded forms of the Takayanagi isostress blending law (Takayanagi *et al.*, 1963). The model provides a tool for estimating multi-layered gel stiffness, when bulging does not occur during compression, and may be applicable to other food systems that behave similarly (Ben-Zion & Nussinovitch, 1996). Another report deals with the compressive force–deformation relationships of multi-layered hydrocolloid gels composed of different combinations of agar, xanthan, carrageenan and konjak mannan and four galactomannans, and texturized fruits based on banana, apple, kiwi and strawberry pulps and agar–LBG combinations, adhered via three different gluing techniques. These relationships were calculated from those of the individual layers (Ben-Zion & Nussinovitch, 1997). The gluing techniques consisted of pouring hot hydrocolloid solution on a gelled layer, using melted agar as a glue between already gelled layers, or simultaneously pouring pre-gelled (gum solution before setting) hydrocolloid solutions (Fig. 4.2). Two assumptions were made: that the normal force in the layers is the same, and that the deformations are additive. The effects of lateral stresses were considered negligible. The calculation was performed using a mathematical model previously developed for double-layered curdlan gels (Nussinovitch *et al.*, 1991). The model constants were determined from the behavior of the individual layers. Good agreement was found between experimental and fitted results over a considerable range of strains. Thus the model's applicability to a given gel system was demonstrated, suggesting a very

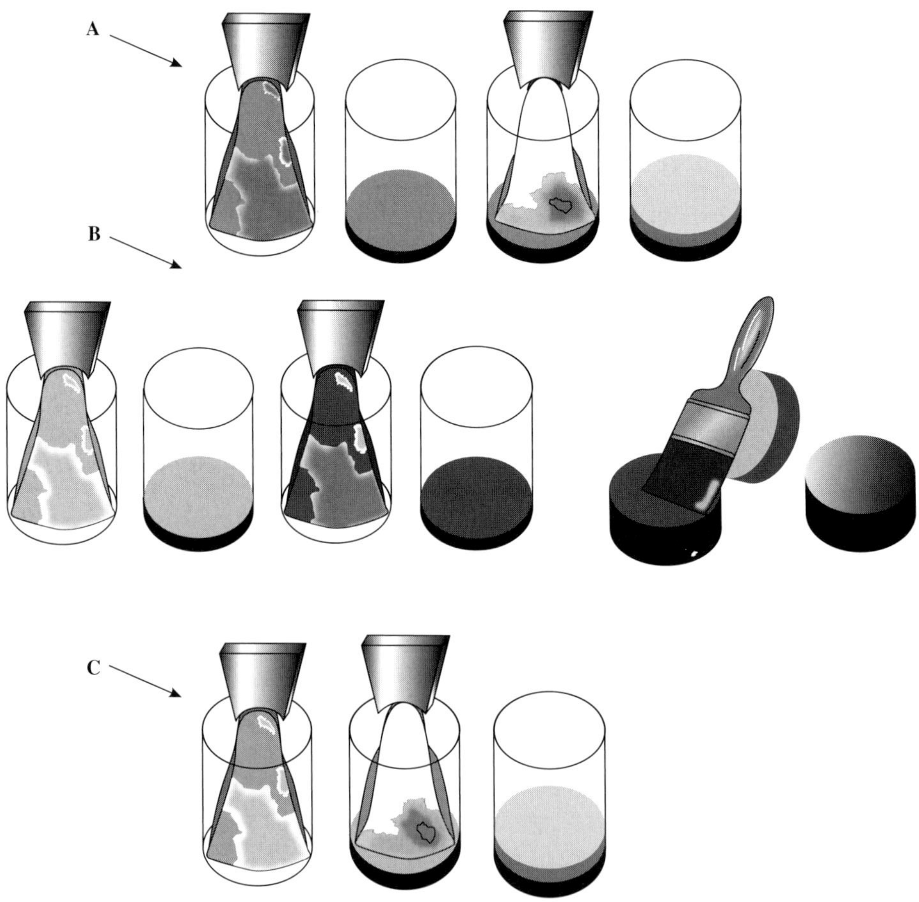

Fig. 4.2 Three different techniques to *glue* the layers of the multi-layered gel together. (A) Pouring hot pre-setting solutions of hydrocolloid or hydrocolloid mixtures on already-gelled layers with identical or different compositions at room temperature will usually (though not always) produce multi-layered gels. (B) Smearing the surface of one layer thoroughly (using a fine brush) with 2% agar solution at 95°C, then pressing the two layers together. (C) Two solutions were prepared simultaneously and were poured together. After a short time, the layers separated, while the solution was still hot. After gelation, a two-layered gel system could easily be observed. (Courtesy of O. Ben-Zion.)

convenient tool for analyzing and predicting the compressive behavior of any number of arrays with different layer combinations (Fig. 4.3) (Ben-Zion & Nussinovitch, 1997; Nussinovitch *et al.*, 1991).

4.3 Other multi-layered edible hydrocolloid products

Agar can be used in confections at concentrations of ~0.3–1.8%. Agar gel was used to manufacture a ready-to-eat, multi-layered, sweetened product (Nussinovitch, 1997). As previously mentioned, the compressive force–deformation relationships of double-layered curdlan gels (made of 2.5% and 3.5% gels) were calculated from those of the components. On the basis of the assumptions that the normal force in the layers is the same, that their deformations are additive, and that the effects of lateral stresses and viscoelasticity can be neglected, the calculation was performed using a mathematical model whose constants were determined from the behavior of the individual layers. The calculated relationships were in agreement with experimental relationships over a considerable range of strains. Failure of the arrays, however, preceded that of the layers and was accompanied by their separation, a phenomenon that is unrelated to the model but limits the strain range of its applicability (Nussinovitch *et al.*, 1991). Curdlan is useful in preparing new types of jelly products. The polymer can be used at a final concentration of 0.4–6.0% in foods. The character of curdlan gels is intermediate between the brittleness of agar gels and the elasticity of gelatin (Kimura *et al.*, 1973). The gel is stable between pH 3.0 and pH 9.5. The gel can absorb sugars quickly and at high concentrations from a syrup, and thus it can be used to prepare sweet jellies (Harada, 1979). The gelling properties of curdlan indicate that it can partially or completely replace polymers such as agar, gelatin or carrageenan. Curdlan forms high-set gels when its aqueous suspension is heated

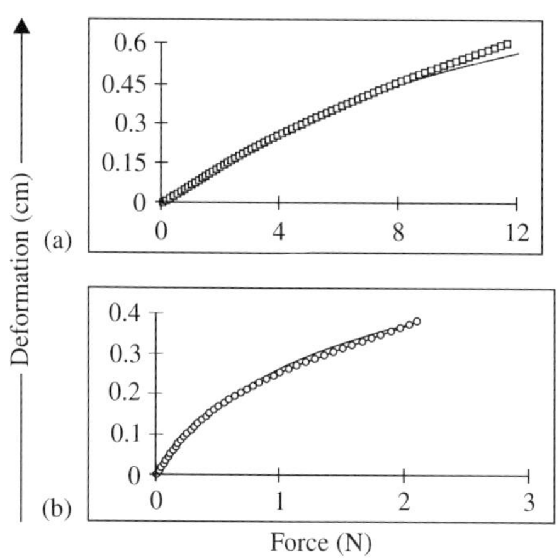

Fig. 4.3 Deformation versus force relationships of (a) a four-layered texturized fruit and (b) a four-layered gel. Both multi-layered gels were constructed by pouring hot hydrocolloid solutions on already-gelled layers. Symbols denote experimental data and the solid lines represent the model's prediction. (From Ben-Zion & Nussinovitch, 1997, with permission from Elsevier Science.)

to about 70–100°C, whereas it forms a low-set, heat-reversible gel upon cooling after heating to ~60 or 65°C. A canned, multiple-layer jelly can be prepared containing both the high-set and low-set gels. This product has been extensively evaluated as a new gel type (Harada, 1979).

Other products can be manufactured using this double-gel-set procedure. Kamaboko, a gelled seafood product from frozen surimi, has distinctive textural properties. Characterization of these properties, using an integrated approach to rheological studies, was accomplished by means of an instrumental texture profile analysis and evaluation of resultant stress–strain relationships. The material had near-ideal area expansion, even at compressions of 60%, while retaining its highly elastic texture. The product did not yield, up to 80% compression. Hardness of the kamaboko at 80% compression was characterized by a local maximum at 37.5°C which may have been related to the processing temperature of the initial surimi gel used in the double-gel-set procedure. Evaluation of stress–strain relationships confirmed the incompressible nature of the gel and showed relatively slight variations between the Young's and deformability moduli. The elastic limit of the kamaboko increased significantly as temperature increased from 25 to 50°C (Konstance, 1991).

4.4 Layered cellular solids

Chapter 8 is devoted in part to a description of the different methods of manufacturing hydrocolloid dry products. When a flat layered array of different cellular materials, each having a different (or the same) thickness, is compressed uniaxially, its cross-sectional area, like that of its individual layers, can be assumed to be practically unchanged (Peleg, 1997). Consequently, the stress along the array can be considered the same in all the layers while the total deformation may be represented as the sum of the deformations of each layer (Fig. 4.4). Expressed mathematically,

$$\sigma_{\text{total}} = \sigma_i \tag{3}$$

where σ_{total} is the array's stress and σ_i is the stress in an individual layer, i, and

$$\varepsilon_{\text{total}} = \left(\frac{1}{H_{0\,\text{total}}}\right) \sum H_{0i}\varepsilon_i(\sigma) \tag{4}$$

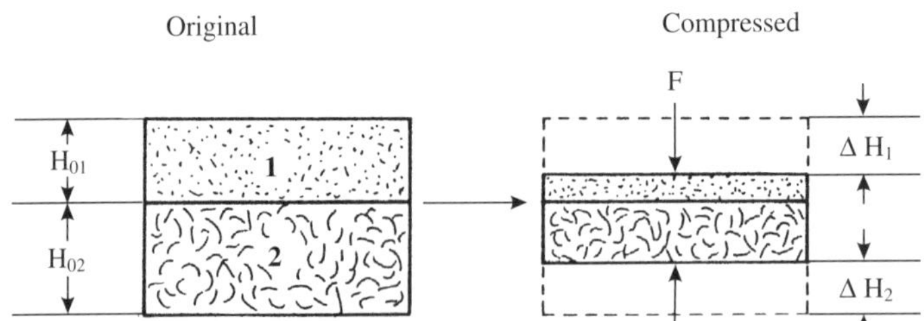

Fig. 4.4 Geometry of a uniaxially compressed double-layered array. (From Swyngedau & Peleg, 1992, with permission from American Association of Cereal Chemists.)

where $\varepsilon_{\text{total}}$ is the array's strain, H_{0i} is the individual layer's thickness, and ε_i is its strain as a function of the stress. The array's initial overall thickness is the sum of that of the individual layers, i.e.

$$H_{0\,\text{total}} = \sum H_{0i} \tag{5}$$

The use of equations as deformability models (see Chapter 8) for cellular solids is especially convenient in that they can be used to express the strain as an explicit algebraic function of the stress $\varepsilon(\sigma)$ (Peleg, 1997). Inserting the terms $\varepsilon_i(\sigma)$ and corresponding H_{0i} into Equation 6 allows a calculation of the stress–strain relationships of any layered array of sponges, as long as the assumption that the cross-sectional area remains practically unchanged remains valid (Swyngedau & Peleg, 1992; Swyngedau *et al.*, 1991a, b).

$$\sigma(\varepsilon) = C_1'\varepsilon^{n_1'} + C_2'\varepsilon^{n_2'} \quad (n_1<1, n_2>1) \tag{6}$$

The difficulty with deformability models is that a determination of their constants by non-linear regression requires accurate guesses to serve as initial values (Peleg, 1997). This difficulty can be eliminated if Equation 6 or a polynomial model is used instead. Arriving at close-enough initial values when Equation 6 is used is fairly easy because it is known that $n_1<1$ and $n_2>1$.

Constants of a polynomial model, such as

$$\sigma(\varepsilon) = C_1\varepsilon + C_2\varepsilon^2 + C_3\varepsilon^3 + C_4\varepsilon^4 \tag{7}$$

can be determined by a generalized linear regression computer program with no need for any initial guesswork (Peleg, 1997).

Once all the $\sigma_i(\varepsilon)$, expressed in terms of any model, and the H_{0i}s are known, the stress–strain relationship of the array can be calculated numerically using standard equation-solving software. All that is needed is to find the root of each σ_i for any desired level of stress or series of stress levels, i.e.

$$\varepsilon_i(\sigma) = \text{root}[\sigma - \sigma_i(\varepsilon) = 0, \varepsilon] \tag{8}$$

and inserting the solution into Equation 6 to generate the array's stress–strain relationships (Peleg, 1993). Once the program is in place, one can change the values of H_{0i} to produce the stress–strain relationship of any desirable array, i.e.

$$\varepsilon_{\text{total}} = \left(\frac{1}{H_{0\,\text{total}}}\right) \sum H_{0i}\text{root}[\sigma - \sigma_i(\varepsilon) = 0, \varepsilon] \tag{9}$$

Although the calculation is made by finding the strain that corresponds to any particular stress, creating the more conventional relationship or plot of σ versus $\varepsilon_{\text{total}}$ is a trivial task (Peleg, 1997).

Micro-computers can be used to analyze or design layered arrays of cellular solids or compressed layers, due to the availability of suitable software which enables rapid calculations. The solution does not require a strain expression as an explicit algebraic function of stress. Therefore, the computer programs can be used with a variety of mixed-model combinations, irrespective of the mathematical structure of the model. All these methods were developed assuming no practical changes in the cross-sectional area of the systems. Another implementation involves closed-cell cellular solids exhibiting small expansions, under large compressive strains (Peleg, 1993).

4.5 Biotechnological uses of multi-layered gels

Biotechnological applications of various multi-layered natural and synthetic gels are not used to a great extent in the food industry or in the development of foods for the future. Here we give just a few examples of their future potential for this industry. Encapsulation of reagents in a multi-layered hydrogel formulation and the dependence of their release on the layered structure have been described for non-food applications (Moriyama *et al.*, 1999). More food-oriented applications might be related to fibers with single or double gel layers for the production of alcohol, for example.

Worldwide, alcoholic beverages account for considerably more than half the gross value of fermented products. Overall, per capita consumption of alcohol is on the rise. The number of alcoholic beverages produced is vast. There are three basic types, defined primarily by means of production: beers, wines and ciders, and distilled products (Sutherland *et al.*, 1986). During alcohol fermentation, immobilization of yeast cells within double-layered gel fibers leads to reduced production of α-acetolactate (diacetyl precursor), relative to immobilization in single-layered gel beads. Reduction in this diacetyl precursor may be due to the more pronounced anaerobic conditions inside the gel fibers (Shindo *et al.*, 1993). Double gel layers can be used in immobilization processes. They are defined and their properties described in Chapter 6 (see also Bucke, 1983; Mattiasson, 1983). There are several techniques via which cells can be immobilized, including adsorption to neutral or charged supports, flocculation, entrapment by natural or synthetic polymers, covalent coupling and containment. In general, desirable features for immobilized cell preparations are high biocatalytic activity, long-term stability of the biocatalyst, possibility of regenerating the biocatalyst, low loss of activity during immobilization, low leakage of cells, non-compressible particles, high resistance to abrasion, resistance to microbial degradation, low diffusional limitations, spherical shape, high surface area, appropriate density for the reactor type, technique simplicity, inexpensive support materials, and non-toxicity of those materials (Nussinovitch, 1997; Tampion & Tampion, 1987).

Another report described the advantages of using bacteria [(Citr+) *Lactococcus lactis* subsp. *lactis* 3022] immobilized in double gel layers of calcium alginate fibers for the production of diacetyl (important carbonyl compound formed during fermentation; the flavor attributed to this compound has been described as *buttery*, *honey*- or *toffee-like*, or *butterscotch*), versus immobilization in calcium alginate single-layer fine fibers. In the past, it was shown that lactic acid bacteria are antagonistic toward many microorganisms due to their competition for nutrients and their production of compounds such as hydrogen peroxide, acetic acid, lactic acid and diacetyl. In addition, citric acid fermentation in conjunction with lactic acid fermentation produces diacetyl and other flavor and aroma compounds in dairy products (Doores, 1990). Diacetyl production by this bacteria was threefold the yield in the presence of catalase. Even higher diacetyl production was observed when a double gel layer of calcium alginate fibers was used holding lactic acid bacteria (outer layer) and homogenized bovine liver as the growth substrate (inner layer) (Ochi *et al.*, 1991).

The advantage of double-layered calcium alginate gel fibers is reflected in their ability to eliminate cell leakage during production. This was demonstrated in an immobilized preparation of *Wasabia japonica* used for chitinase production. Chitinase production by the immobilized cells was higher than that detected with freely suspended cells under the same conditions. Aeration with pure oxygen resulted in increased chitinase production, about fivefold that of the freely suspended cells (Tanaka *et al.*, 1996a). To maintain chitinase production above 80% of its maximal rate, it was necessary to keep its concentration in the broth below $2\,U\,ml^{-1}$. A specialized production arrangement involving double-layered gel fibers coupled to a chitin column permitted stable production of chitinase for 40 days (Tanaka *et al.*, 1996b). Other biotechnological processes involve encapsulation of cells, enzymes, proteins, adsorbents and magnetic materials for food and non-food applications (see Chapter 5) (Bader *et al.*, 1995; Chang, 1993; Grant *et al.*, 2000). Cells can be cryopreserved to maintain their viability and for other purposes. Most of the information on such procedures is to be culled from the microbiological or medical literature (Koebe *et al.*, 1999).

4.6 Techniques to evaluate properties of multi-layered products

The techniques and software that are and will be used in evaluating multi-layered products stem from many branches of science. Software for the optical characterization of thin films using spectrophotometric and/or ellipsometric measurements have been developed. The program enables the analysis of a wide range of multi-layered structures, with respect to the composition, micro-structure or thickness of any of the layers (Leinfellner *et al.*, 2000). X-ray diffraction is a powerful tool for investigating residual stress states in micro-systems, because it enables the non-destructive testing of materials and components, and the measurement of very small volumes. Due to the small thicknesses of the individual layers in multi-layered micro-systems, in many cases the X-rays penetrate deeply into the samples, and diffraction patterns occur in layers of different materials. New detection systems such as area detectors – their evaluation pending – can be used successfully (Kampfe, 2000).

Today, multi-layered food products based on hydrocolloids are rare. However, the apparent trend towards these foods cannot be ignored. Therefore, methods from different, more technical areas need to be adopted for the development and analysis of these current and future products. A good example might be the previously proposed analytical method for studying the propagation of transient elastic waves in multi-layered media (Lee & Ma, 1998). A dynamic displacement response can be obtained by setting up experimental conditions that simulate planar stress conditions for a layered half-space. Experimental and theoretical solutions have gone hand in hand and the method can also be applied to a three-dimensional space (Lee & Ma, 1998). Another method discusses the application of the Rayleigh–Ritz method (a variational method for non-equilibrium statistical dynamics), for analyzing multi-layered plates with residual stresses for membrane and bending deformation. The advantage of this method is its simplicity and ease of application to simple sample shapes (Lee & Kim, 1999). Other methods to be tested for their suitability to such complicated products are ultra-micro-hardness, adhesion and residual-stress analyses, especially in systems where one layer is very thin, i.e. it resembles a film that coats a surface (Tavares *et al.*, 1999).

4.7 References

Bader, A., E. Knop, K. Boker, N. Fruhauf, W. Schuttler, K. Oldhafer, R. Burkhard, R. Pichlmayr & K.F. Sewing (1995) A novel bioreactor design for in-vitro reconstruction of in-vivo liver characteristics. *Artif. Organs*, **19**(4), 368–374.

Ben-Zion, O. & A. Nussinovitch (1995) Calculating the compressive deformabilities of multi-layered gels and texturized fruits glued together by three different adhesion techniques. The Eight International Conference and Industrial Exhibition on Gums and Stabilisers for the Food Industry. 1–14 July, The North East Wales Institute, Cartrefle College, Wrexham, Clwyd, UK.

Ben-Zion, O. & A. Nussinovitch (1996) Predicting the deformability modulus of multi-layered texturized fruits and gels. *Lebensm.-Wiss. u.-Technol.*, **29**, 129–134.

Ben-Zion, O. & A. Nussinovitch (1997) A prediction of the compressive deformabilities of multilayered gels and texturized fruit, glued together by three different adhesion techniques. *Food Hydrocolloids*, **11**(3), 253–260.

Bucke, C. (1983) Immobilized cells. *Philosophical Transactions of the Royal Society, London, Series B*, **300**, 369–389.

Chang, T.M.S. (1993) Living cells and microorganisms immobilized by microencapsulation inside artificial cells. In: (Goosen, M.F.A., Ed.), *Fundamentals of Animal Cell Encapsulation and Immobilization*. CRC Press, Boca Raton, FL, p. 184.

Doores, S. (1990) pH control agents and acidulants. In: (Larry Branen, A., P. Michael Davidson & S. Salminen, Eds.), *Food Additives*. Marcel Dekker, Inc., New York and Basel, chapter 13, pp. 477–510.

Grant, M.H., K. Anderson, G. McKay, M. Wills, C. Henderson & C. MacDonald (2000) Manipulation of the phenotype of immortalised rat hepatocytes by different culture configurations and by dimethyl sulphoxide. *Hum. Exp. Toxicol.*, **19**(5), 309–317.

Harada, T. (1979) Curdlan: a gel-forming β-1,3 glucan. In: (Blanshard and Mitchell, Ed.), *Polysaccharides in Food*. Butterworths, London, pp. 283–300.

Ikeda, T., S. Moritaka, S. Sugiura & T. Umeki (1976) Method for preparing jelly foods. United States Patent 3,969,536.

Kampfe, B. (2000) Investigation of residual stresses in microsystems using X-ray diffraction. *Materials, Science and Engineering A – Structural Material, Properties, Microstructure and Processing*, **288**(2), 119–125.

Kimura, H., S. Moritaka & M. Misaki (1973) Polyscaccharide 13140: a new thermo-gelable polysaccharide. *J. Food Sci.*, **38**, 668–670.

Koebe, H.G., B. Muhling, C.J. Deglmann & F.W. Schildberg (1999) Cryopreserved porcine hepatocyte cultures. *Chem. Biol. Interact.*, **121**(1), 99–115.

Konstance, R.P. (1991) Axial compression properties of kamaboko. *J. Food Sci.*, **56**(5), 1287–1291.

Lee, B.C. & E.S. Kim (1999) A simple and efficient method of analyzing mechanical behaviors of multi-layered orthotropic plates in rectangular shape. *J. Micromechanics Microengineering*, **9**(4), 385–393.

Lee, G.S. & C.C. Ma (1998) Transient elastic waves propagating in a multi-layered medium subjected to in-plane dynamic loadings. *Proceedings of the Royal Society of London Series A – Mathematical, Physical and Engineering Sciences*, **456**, 1355–1374.

Leinfellner, N., J. Ferre-Borrull & S. Bosch (2000) A software for optical characterization of thin films for microelectronic applications. *Microelectronics Reliability*, **40**(4–5), 873–875.

Mattiasson, B. (1983) *Immobilized Cells and Organelles*, vols. 1 & 2. CRC Press, Boca Raton, FL.

Moriyama, K., T. Ooya & N. Yui (1999) Pulsatile peptide release from multi-layered hydrogel formulations consisting of poly(ethyleneglycol)-grafted and ungrafted dextrans. *J. Biomater. Sci. Polymer Edition*, **10**(12), 1251–1264.

Nussinovitch, A. (1997) In: *Hydrocolloid Applications: Gum Technology in the Food and Other Industries*. Blackie Academic & Professional, London, pp. 14–15.

Nussinovitch, A., S.J. Lee, G. Kaletunc & M. Peleg (1991) Model for calculating the compressive deformability of double-layered curdlan gels. *Biotechnol. Prog.*, **7**, 272–274.

Ochi, H., M. Takahashi, T. Kaneko, H. Suzuki & H. Tanaka (1991) Diacetyl production by co-immobilized citrate-positive *Lactococcus lactis* subsp *lactis* 3022 and homogenized bovine liver in alginate fibers with double gel layers. *Biotechnol. Lett.*, **13**(7), 505–510.

Peleg, M. (1993) Calculation of the compressive stress-strain relationships of layered arrays of cellular solids using equation-solving computer software. *J. Cellular Plastics*, **29**, 285–293.

Peleg, M. (1997) Review: Mechanical properties of dry cellular solid foods. *Food Sci. Technol. Int.*, **3**, 227–240.

Shindo, S., H. Sahara, S. Koshino & H. Tanaka (1993) Control of diacetyl precursor (alpha-acetolactate) formation during alcohol fermentation with yeast-cells immobilized in alginate fibers with double gel layers. *J. Fermentation and Bioengineering*, **76**(3), 199–202.

Sutherland, J.P., A.H. Varnam & M.G. Evans (1986) In: *A Color Atlas of Food Quality Control*. A Wolfe Science Book, Weert, The Netherlands, p. 243.

Swyngedau, S. & M. Peleg (1992) Characterization and prediction of the stress–strain relationship of layered arrays of spongy baked goods. *Cereal Chem.*, **69**, 217–221.

Swyngedau, S., A. Nussinovitch & M. Peleg (1991a) Models for the compressibility of layered polymeric sponges. *Polym. Eng. Sci.*, **31**(2), 140–144.

Swyngedau, S., A. Nussinovitch, I. Roy, M. Peleg & V. Huang (1991b) Comparison of four models for the compressibility of breads and plastic foams. *J. Food Sci.*, **56**, 756–759.

Takayanagi, M., H. Harima & Y. Iwata (1963) Viscoelastic behaviour of polymer blends and its comparison with model experiments. *Mem. Fac. Eng. Kyushu Univ.*, **23**, 1–13.

Tampion, J. & M.D. Tampion (1987) *Immobilized Cells: Principles and Applications*. Cambridge University Press, Cambridge.

Tanaka, H., Y. Kaneko, H. Aoyagi, Y. Yamamoto & Y. Fukunaga (1996a) Efficient production of chitinase by immobilized *Wasabia japonica* cells in double-layered gel fibers. *J. Fermentation Bioeng.*, **81**(3), 220–225.

Tanaka, H., T. Yamashita, H. Aoyagi, Y. Yamamoto & Y. Fukunaga (1996b) Efficient production of chitinase by *Wasabia Japonica* protoplasts immobilized in double-layered gel fibers. *J. Fermentation Bioeng.*, **81**(5), 394–399.

Tavares, C.J., L. Rebouta, M. Andritschky & S. Ramos (1999) Mechanical characterisation of TiN/ZrN multi-layered coatings. *J. Mater. Processing Technol.*, **93**, 177–183.

Chapter 5
Hydrocolloids in Flavor Encapsulation

5.1 Introduction

Encapsulation enables the creation of a dry, free-flowing powdered flavor (Reineccius, 1991). A coating protects the flavoring from interaction with the food, inhibits oxidation and can enable controlled flavor release. A variety of processes can be used to encapsulate the flavoring within the film, with the latter's properties being dependent upon processing as well as composition. Of the many processes described for flavor encapsulation, spray-drying and extrusion are the most commercially advantageous (Ré, 1998; Reineccius, 1989).

Many reports and reviews on food encapsulation can be found in the literature. King (1995) provides a brief definition of the terms and nomenclature used in the field and discusses the use of hydrocolloids in encapsulation. Another report describes the use of encapsulated flavoring to preserve food flavor (Anon, 1994), and the pros and cons of different encapsulation systems have been previously described (Taylor, 1983). Developments in flavor-encapsulation systems for foods have been discussed (Mothes, 1997), and a descriptive review of micro-encapsulation of food ingredients can also be found elsewhere (Tuley, 1996).

5.2 Spray-drying for flavor encapsulation

More than 90% of the encapsulated flavoring on the market today is produced by spray-drying. This technique has been widely used for drying heat-sensitive foods, pharmaceuticals, and other substances, because of the solvent's rapid evaporation from the droplets. Although most often considered a dehydration process, spray-drying can also be used as an encapsulation method when it entraps *active* material within a protective matrix, which is essentially inert to the material being encapsulated. Compared to the other conventional micro-encapsulation techniques, it offers the attractive advantages of producing micro-capsules in a relatively simple, continuous processing operation.

Spray-drying generally involves producing an emulsion of the flavoring in an encapsulation matrix (at a ratio of about 1:4, respectively, on a dry weight basis), taking into consideration that a minimal amount of water, preferably 40% (wet weight), needs to be included in the formulation. Homogenization is used to prepare the emulsion with a particle size averaging ~1 μm. The emulsion is fed into a spray-dryer, where it passes through atomization into hot air streams where evaporative cooling brings the inlet air temperature down from 200–325°C to 80–90°C (Nussinovitch, 1997). Complete and uniform atomization is necessary for successful drying, and one of the following types of atomizers is used: centrifugal atomizer (liquid is fed to the center of a rotating bowl and droplets are flung from the edge of the bowl to form a uniform spray), pressure-nozzle atomizer (where liquid is forced at high pressure through a small aperture; grooves on the inside of the nozzle cause a cone-shaped spray, thus making use of the full volume of the drying chamber), and a two-fluid nozzle atomizer (compressed air creates turbulence which atomizes the liquid; the operating pressure is lower than the nozzle pressure, but a wider range of droplet sizes is produced) (Fellows, 1990).

During drying, the temperature of the flavoring particle never exceeds that of the exit air. The matrix materials need to be water-soluble, making hydrocolloids a good choice. The water solubility of the matrix is important because many flavoring materials are designed to be released by contact with water. Since continuous pumping and atomization of the feed material is essential for processing, two main prerequisites are low viscosity and a high concentration of solids (50–70%). The emulsion should be kept stable until its atomization (Nussinovitch, 1997). The encapsulation matrix should not become sticky at high temperatures, so that good yield can be achieved; after manufacturing, the product should not be hygroscopic. Of course, once all of these requirements have been met, the encapsulated flavoring needs to have retained its volatile flavor constituents; the capsule should be capable of resisting evaporation and degradation during storage and of providing appropriate release of the finished product. All of these requirements limit the use of matrix materials. Maltodextrins, corn syrup solids, modified starches and acacia gum are examples of suitable matrix builders and encapsulators (Nussinovitch, 1997).

A special spray-drying method – *Leafflash* – was developed, wherein very hot air flows at a very high velocity, allowing drying of highly viscous liquids. An example could be a mixture of two volatile products (Bhandari *et al.*, 1992). Citral ($C_{10}H_{16}O$, synthetic flavoring, pale-yellow liquid having strong, lemon-like odor, a characteristic bittersweet taste, originally found in lemongrass oil and other sources) and linalyl acetate ($C_{12}H_{20}O_2$, colorless synthetic flavoring with a bergamot–lavender odor and a persistent sweet, acrid taste), in proportions of 80:20 (w/w), and blends of maltodextrin and gum arabic were included within the formulations. Emulsions were produced, atomized and dried at inlet air temperatures of 300–400°C. No adverse effect on chemical properties was observed. The use of maltodextrin (less expensive) for delicate flavor encapsulation, instead of gum arabic, is possible in an emulsion containing 60% total solids (Bhandari *et al.*, 1992).

5.3 Extrusion processes for flavor encapsulation

Food ingredients can be encapsulated by extrusion, a relatively new process compared to spray-drying. Extrusion here does not resemble that used for cooking or texturization of cereals. It is a relatively low-temperature entrapping method, which involves forcing a core material dispersed in a molten carbohydrate mass through a series of dies into a bath of dehydrating liquid. Pressure is kept under 100 psi and temperature under 115°C (Pegg & Shahidi, 1999). Upon contact with the liquid, the coating material, which forms the encapsulating matrix, hardens and entraps the core material. For dehydration and hardening, isopropyl alcohol is commonly used. Small pieces are achieved by breaking the extruded strands and further drying and sizing (Pegg & Shahidi, 1999).

Encapsulated food ingredients that have undergone emulsification include orange peel oil in a molten dextrose mass, fruit essences, volatile substances and orange juice solids. The process currently used in today's flavor industry resulted from improvements made in 1957 by Swisher. He added orange peel oil containing antioxidant and dispersing agent to an aqueous melt of corn syrup solids (42 *dextrose equivalent* (DE)) and glycerine. In some processes starch may be degraded by a dual acid–enzyme conversion catalyzed system. The degree of conversion is defined by the DE, which is the percentage of the theoretical maximum hydrolysis which has actually occurred. Agitation of the syrup mixture under a nitrogen blanket formed an oxygen-free emulsion. This emulsion was forced through a die into mineral oil, followed by rapid cooling, extrusion and solidification (Swisher, 1957a, b). Since Swisher's patents, most developments have focused on studying different materials to serve as the encapsulating matrix. A few examples are replacement of high DE corn syrup solids with a combination of sucrose and maltodextrin (Beck, 1972) and use of emulsifying starches in the encapsulation matrix to increase the loading capacity to 40% flavoring (Barnes & Steinke, 1987). Other patents deal with optimization of the extrusion process, for example the finding that high-load products (>22%) have an optimum cooking temperature of about 123°C (Miller & Mutka, 1985, 1986). Overall, extrusion provides a true encapsulation process in the sense that the core material is fully embedded within the matrix wall. Its drawback lies in its high cost relative to spray-drying, since it delivers less flavor per unit weight and flavors must be able to tolerate 110–120°C for substantial periods of time (Pegg & Shahidi, 1999).

5.4 Film performance in flavor encapsulation

For food ingredient encapsulation, the selection of an appropriate coating material (referred to as shell, wall material or encapsulating agent) is the most important (Pegg & Shahidi, 1999). Coating substances are film-formers. They can be selected from natural or synthetic polymers. Many characteristics are required and no single coating material can support all of these demands. Encapsulation allows separation of reactive ingredients from their environment until their desired release (Pegg & Shahidi, 1999). In formulated foods, a substance can be encapsulated such that it will not diffuse through during processing, but only later, during consumption. Slow release is necessary in products such as chewing gums or baked and microwaved products. Many other examples exist. They include acidulants, meat-processing aids and dough conditioners. The coating protects the core material from oxygen, light, other food ingredients, external agents and moisture. However, it also participates in the controlled release, which in turn depends on the capsule's geometry, type and wall material, as well as on solvent effects, particle degradation or fracture and diffusion

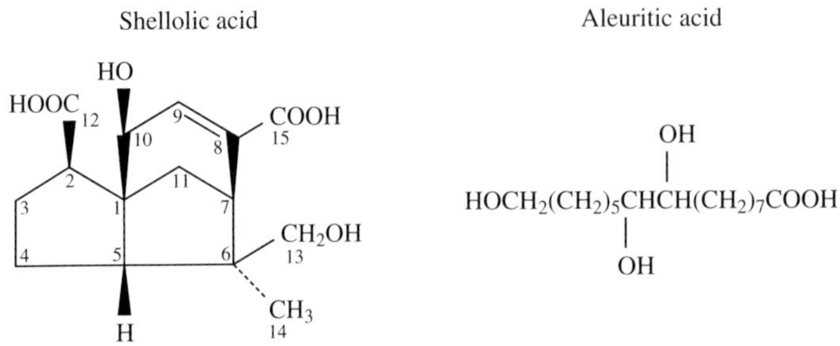

Fig. 5.1 The major chemical constituents of shellac.

(Pegg & Shahidi, 1999). Control properties can also be established by using secondary coatings [with, e.g. fats, oils, edible shellacs (Fig. 5.1), modified cellulose] around an already encapsulated flavoring. Since this coating dilutes flavor strength and adds to the cost, better techniques are constantly being sought (Nussinovitch, 1997; Veronese, 1974).

Skin-forming materials, classified as *heat-sensitive products*, will form a skin under any drying conditions provided that the initial solid concentration is high enough. To investigate flavor retention, ethanol is sometimes chosen as a flavor simulant. Details of two cases in which ethanol was used for this purpose are briefly described here, due to the important research principal involved. For example, retention of ethanol was studied in three heat-sensitive encapsulants, i.e. gum arabic and gelatin (Bloom numbers 60 and 150). Single droplets were dried in a horizontal wind tunnel and differences in ethanol concentration and crust structure were determined. The stronger gel produced by gelatin (Bloom number 150) enabled it to retain the most ethanol. Gum arabic did not enhance ethanol retention. Surface cracks on droplets caused loss of volatiles. Results demonstrated that the retention of volatiles during convective drying is a function of selective diffusion (Hassan *et al.*, 1996).

Skin (crust) formation during encapsulation of food flavors can be investigated by using native rice starch, wheat starch or dextrin as skin-forming materials and ethanol as a simulated flavor (Sayed *et al.*, 1996). Ethanol concentration was estimated by gas–liquid chromatography and its retention was defined arbitrarily as contents of the capsule (droplet) after 10 min of drying. With rice starch, no droplet expansion (which enhanced loss of volatiles) was observed. Wheat starch and dextrin also provided no droplet expansion and demonstrated increased flavor retention as a function of time and the mechanism of selective diffusion through the formed skin (Sayed *et al.*, 1996).

5.5 Materials for flavor encapsulation

Most flavors are encapsulated in water-soluble polymers: acacia gum, modified starches, maltodextrin and corn syrup solids, with acacia gum being the traditional choice (Thevenet, 1986). Production of the gum (Fig. 5.2), and its physical and chemical properties are described elsewhere (Nussinovitch, 1997; Phillips & Williams, 2000). Gum arabic from *Acacia senegal* contains ~3.8% ash, 0.34% nitrogen, 0.24%

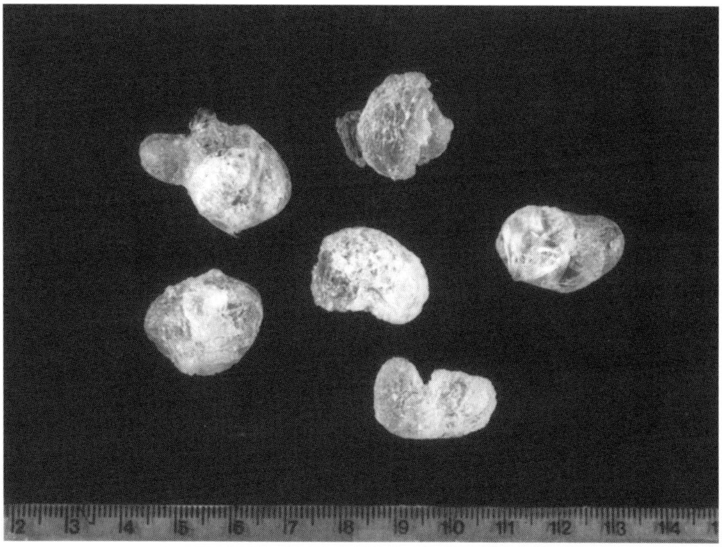

Fig. 5.2 *Tear drops* of gum arabic, the dried, gummy exudate obtained from acacia trees.

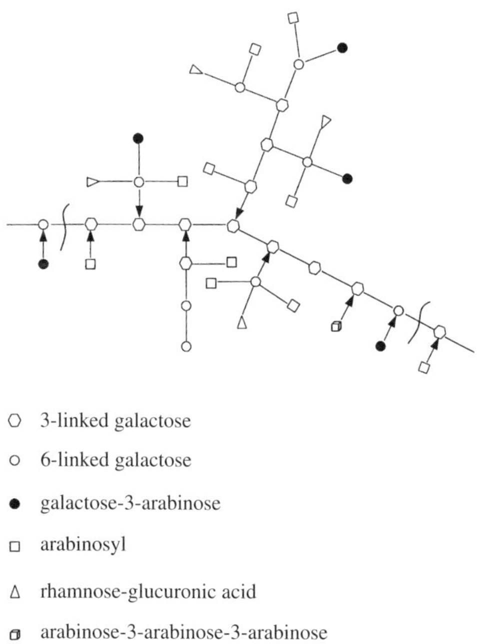

Fig. 5.3 Possible structure of the carbohydrate component of gum from *Acacia senegal.*

methoxyl, 17% uronic acid, and the following sugar constituents after hydrolysis: 45% galactose, 24% arabinose, 13% rhamnose, 16% glucuronic acid and 15% 4-*O*-methyl glucuronic acid (Anderson *et al.*, 1990). The gum is a slightly acidic complex polysaccharide (Fig. 5.3) produced as a mixture of calcium, magnesium and potassium salts. It has a molecular mass of ~580 000 Da. Gums from different sources exhibit large differences in amino acid composition, uronic acid content and molecular weight. As the main product for food applications, gum arabic from *A. senegal* is consistent in quality. Three principal fractions have been identified by hydrophobic-affinity chromatography: a low-molecular-weight arabinogalactan (AG), a very high-molecular-weight AG-protein complex (AGP) and a low-molecular-weight glycoprotein (GI). These components represent 88%, 10% and 1% of the molecule, respectively, and they contain 20%, 50% and 30% polypeptides, respectively. The protein is located on the outside of the AGP unit. The overall conformation of the gum arabic molecule is described by the *wattle blossom* model in which approximately five bulky AG blocks, of ~200 000 Da each, are arranged along the GI polypeptide chain which may contain up to 1600 amino acid residues (Connolly *et al.*, 1987; Imeson, 1992; Phillips & Williams, 2000).

On the one hand, starch can entrap flavor molecules and produce stable complexes. On the other, hydrolysates derived from it confer virtually no emulsification properties to an encapsulated compound (Pegg & Shahidi, 1999). This disadvantage creates two problems: poor flavor retention and rapid flavor separation from the product. To overcome this drawback, starches can be chemically modified to change their functionality. For example, the U.S. Food and Drug Administration approved the treatment of starch with 1-octenylsuccinic anhydride to form a modified starch including hydrophobic and hydrophilic groups (Pegg & Shahidi, 1999). Octenyl succinate can be used at no more than 0.02 degrees of substitution on the starch polymer. After substitution, the modified starches serve as excellent emulsifiers and, at a high solids content (40–50%), still exhibit low viscosity. If starches are to be treated with acid or an acid–enzyme combination, then corn syrup solids or maltodextrins can be used. Products with a DE of less than 20 are classified as maltodextrins. Products having a DE equal to or higher than 20 are corn syrup solids. Both are used at levels which produce high viscosities, to maintain a temporary emulsion but not confer emulsion stability on the finished product. Maltodextrins and corn syrup solids are inexpensive relative to other hydrocolloids, and can also be used as fillers (Trubiano & Lacourse, 1986).

Another report deals with sodium starch–octenyl succinates as special food starches characterized by low viscosity at high solids concentrations (40–50%) and the ability to form very fine, stable emulsions, two key factors for very effective encapsulation of flavor oils by spray-drying (Trubiano, 1995). The results show the superior performance of these starch products over other encapsulating agents or carriers, such as powdered gum arabic and maltodextrins. Parameters used to evaluate these unique encapsulating agents include emulsion particle size and stability, level of orange oil that is actually retained in the spray-dried powders, exposed surface oil, stability of the flavor toward evaporation, oxidation resistance of the encapsulated powders and spray-drying rates (Trubiano, 1995). Gum arabic is the most commonly used encapsulation coating material via spray-drying. It is a natural emulsifier and is used as a flavor fixative in the production of powdered aroma concentrates (Pegg & Shahidi, 1999). Compared to maltodextrins, gum arabic (acacia) gives superior aroma retention during drying and storage. Maltodextrin and gum acacia mixtures are successfully used for encapsulation of flavors that are stable against oxidation (Bhandari *et al.*, 1992). Another past reference reviewed other materials' properties for micro-encapsulation, such as karaya gum, gum tragacanth, acacia gums, gum arabic (structure, viscosity, acidity, surface tension), and gum arabic as an emulsifier and stabilizer (liquid emulsions, encapsulation by spray-drying, and micro-encapsulation by coacervation) (Balke, 1984).

Alginates are powerful thickening, stabilizing and gel-forming agents and in addition to their traditional uses, they can also serve as encapsulating agents. A few reports discuss this usage. Flavor-neutral micro-capsules are made from a combination of indigestible materials, e.g. oligo- or polysaccharides such as alginate, in combination with digestible materials. They may be used as carriers for food supplements, e.g. minerals or unpleasant tasting oils, or for pharmaceuticals (Heinrich *et al.*, 1998). A process for micro-encapsulation of organic materials (especially hydrophilic substances in the crystalline state) in particles with walls formed from a cross-linked polymer is based on mixing an aqueous colloidal solution of a cationic polymer, such as gelatin below its isoelectric point, with an aqueous colloidal solution of an anionic polymer of high ionic strength, together with the organic material to be micro-encapsulated, to form an oil-in-water emulsion at a temperature higher than the melting point of the material to be encapsulated. The emulsion is then chilled and cross-linked, before neutralization and recovery of the micro-particles. This process may be applied to food additives (Benoit *et al.*, 1997). Alginate was also reported to produce micro-spheres by forming an aqueous slurry containing an immobilizing agent (e.g. gum) and a gelling agent; the slurry is then put into contact with an oil-soluble organic acid, to convert droplets to micro-spheres. Food applications include encapsulation of food flavorants (Lencki *et al.*, 1989).

5.6 Flavor oxidation, retention and shelf-life

Decreases in the quality and shelf-life of flavoring materials are due to the oxidation of flavoring agents and the consequent development of off-flavors. Maltodextrins provide varying amounts of protection, depending upon their DE (Anandaraman & Reineccius, 1986): the higher the DE, the better the protection. This poses a problem since higher DE materials tend to be difficult to dry, are very hygroscopic and result in poor flavor retention, leaving open the possibility of achieving an unsuitable product with an excellent shelf-life. Modified food starches are used to achieve encapsulated flavoring with good retention, but poor protection against oxidation (Nussinovitch, 1997).

A preparation's shelf-life is related to the permeability of the encapsulating matrix to oxygen (Baisier & Reineccius, 1989). The rate of molecular diffusion is dependent upon the state of the polymer composing the capsule. At low water activity and temperature, the related polymers are in a glassy state, limiting molecular diffusion; at higher water activity and temperature, the material changes to a rubbery state, permitting molecular diffusion (Levine & Slade, 1989; Nelson & Labuza, 1992). In other words, when the encapsulating film remains in a glassy state, oxygen diffusion is limited or non-existent, whereas it occurs when the film changes to a rubbery state.

It should be noted that this relationship was not found for the oxidation of orange oil which had been spray-dried using maltodextrin as the encapsulating matrix. Nevertheless, as water activity increased, the product was less stable.

Flavor retention during the drying process is essential to the success of a preparation. It should be noted that the more volatile constituents can easily be lost. The volatile organic compounds which are carried away by the drying air need to be removed by scrubbing, and thus poor retention creates complications during processing. In a comparison of the encapsulating efficiencies of acacia gum versus starch carriers (Trubiano & Lacourse, 1986), percent true encapsulated flavor was highest when the low-viscosity starch octenyl succinate was used (29.4% versus 23.9% for acacia gum and 17.4% for dextrin). It should be noted that modified food starches can typically be used at a concentration of 50–55% solids, versus 30–35% solids for acacia gum. A higher level of solids results in better flavor retention (Reineccius, 1989). The poor emulsification properties obtained using maltodextrins result in poor flavor retention during drying.

Flavor retention depends on many factors. In corn-starch-based exudates containing cinnamic aldehyde (C_9H_8O; synthetic flavoring with a pungent spicy note and burning taste; a greenish-yellow liquid, naturally occurring in the essential oils of Ceylon and Madagascar cinnamon leaves), eugenol ($C_{10}H_{12}O_2$; a colorless to slightly yellow liquid having a strong, clove-like odor and spicy, pungent taste), nonanoic acid ($C_9H_{18}O_2$; colorless oily liquid reportedly found in several essential oils, either free or esterified: rose; geranium and oris; having a characteristic fatty, and a corresponding unpleasant taste), and 3-octanone ($C_8H_{16}O$; a colorless mobile liquid with a strong, penetrating, fruity odor reminiscent of lavender), flavor retention was higher in samples stored at –20°C, at lower moisture content and higher initial levels of aroma compounds (Kollengode & Hanna, 1996). Retention of individual artificial food flavor components was dependent on the relationship between in-feed dryer solids and water: 80% retention was observed for gum arabic with 40% in-feed solids, 78 and 64% retention, respectively, for N-Lok (National Starch Corp., Bridgewater, NJ) and Maltrin M-100 (Grain Processing Corp.) with 45% in-feed solids (Reineccius & Bangs, 1985). Another report describes the superior aroma retention of microcapsules, produced of gum arabic (four parts) and containing methyl anthranilate (one part aroma). Very little aroma was lost during storage at humidities below the water mono-layer level, while this loss increased at higher water activities (Rosenberg & Kopelman, 1983). The predominant retention mechanism of labeled ethyl acetate, n-propanol and acetone in freeze-dried gels was claimed to be entrapment in microregions, with a small contribution being due to adsorption (Kayaert *et al.*, 1975). Changing the viscosity of the feed solution to higher than 150 cP can eliminate distortion of feed droplets, yielding very high retention (Kerkhof & Thijssen, 1974). Many aroma compounds are used in a solid state to flavor food products after encapsulation. Retention is dependent on the physicochemical properties of both the volatiles and the carriers, increasing with molecular weight and decreasing with polarity and relative volatility of the aroma compound (Goubet *et al.*, 1998). An increase in aroma goes hand in hand with an increase in the molecular weight of the carrier to an optimum; it then decreases for very high degrees of polymerization.

Carriers in an amorphous state give the highest retention, while considerable loss is connected with carrier collapse and crystallization (Goubet *et al.*, 1998).

5.7 Controlled release of active ingredients

Controlled release of active ingredients is a requisite goal in the art and science of flavor encapsulation. Micro-encapsulation can be tailored to reach this goal. The key indices governing the mechanisms of release involve one or a combination of the following parameters: change in temperature, moisture or pH, shear or pressure application, and addition of surfactants. Encapsulation can protect ingredients (such as antioxidants, enzymes, flavors and preservatives) from temperature, moisture, microorganisms or other components of the food system. Optimal controlled release is achieved when encapsulated ingredients are released at pre-planned sites, times and rates (Pothakamury & Barbosa Canovas, 1995). Gelatin–sucrose gels can be used as a model for flavor release from jellied sweets. The release depends on the melting temperature of the gel, which in turn depends on the concentrations of gelatin and sucrose (Harrison & Hills, 1996). If the melting point occurs at temperatures below mouth temperature, then the release is dependent on the rate of heat diffusion into the gel matrix. In gels with melting points above mouth temperature, the rate-limiting step for flavor release is the diffusion of sucrose from the surface of the gel into the adjacent saliva phase. Practical experimentation verified the agreement between these theoretical explanations and real life (Harrison & Hills, 1996).

5.8 Food applications

Various patents dealing with the encapsulation of food ingredients have been reviewed by Risch (1995) in sections covering controlled release (particularly of sweeteners and flavors from chewing gums), carrier materials, methodology patents (production methods of encapsulated materials), and other patents (e.g. protection of specific ingredients, use of liposomes and coacervation) (Risch, 1995).

5.8.1 Oil and carotenoid encapsulation

Because of their relevance to cosmetics and therapeutics, public interest in essential oils and carotenoid consumption or use is on the rise. Thus micro-encapsulation techniques aimed at these applications are steadily gaining popularity and importance. Lemon oil (obtained by mechanical or manual cold expression of the peel in yields of ~4% based on fruit weight, contains ~90% limonene and other important terpenes) micro-encapsulation was performed using a precipitation technique with oil/β-cyclodextrin at ratios of 3:97, 6:94, 9:91, 12:88 and 15:85 (w/w). At the ratio of 6:94, retention of lemon oil volatiles was at its peak. Maximal inclusion capacity of β-cyclodextrin and powder recovery were achieved at the ratio of 12:88 with a complex including ~10% lemon oil (Bhandari *et al.*, 1998). To reach a potent formulation of encapsulated lemon oil, various hydrocolloids were included into such micro-capsules and their contribution was evaluated. In a comparison of gum arabic, maltodextrin with a DE of 10, lipophilic starch, cornstarch, oxidized-phenylalanine and oxidized-aspartame–glycoamine starches, lipophilic starch was best at oil retention, emulsion stability and load capacity (Bangs & Reineccius, 1990). Gum arabic and sodium caseinate were least effective as orange oil micro-encapsulants on the basis of dynamic headspace analysis results, and whey protein isolate (WPI) and gum arabic were least effective as orange oil micro-encapsulants on the basis of total oil retention and surface oil results (Kim & Morr, 1996).

Emulsified orange essential oils were encapsulated using different matrix-forming agents. Soy protein isolate (SPI)-emulsified orange oil droplets were the most stable during creaming. Gum arabic-emulsified orange oil droplets were the least stable. The sizes of the spray-dried encapsulated particles followed the order of whey protein isolate > SPI > gum arabic. Capsules created with gum arabic exhibited greater shrinkage during drying and had the highest release rate. Those micro-capsules in which proteins were involved had the lowest release rate and were more resistant to oxidation (Young, 1996). Various acacia gums were used as matrix builders, for encapsulated orange oil with or without maltodextrins and commercial modified starch. Acacia gum performed better than modified starch for decreasing oxidation rate but was less efficient at retaining flavors during drying (Reineccius *et al.*, 1995).

One report discussed the reaction of a starch polymer with a lipophilic grouping (on a low-viscosity base starch to allow solids usage in the typical range of gum arabic) by The National Starch & Chemical Corporation to produce a modified starch, called Capsul. It has excellent emulsification properties, sufficient to put citrus oils, vegetable oils, and a wide variety of flavorings into a stable emulsified form for efficient spray-drying (King *et al.*, 1976). A comparison of homogenization tests carried out on various encapsulating agents (corn dextrin, tapioca dextrin, maltodextrin, gum arabic and Capsul) showed that gum arabic and Capsul offer the most stable emulsions, which do not separate prior to spray-drying. The following encapsulation efficiencies were determined: Capsul 93.5%, gum arabic 83.5%, corn dextrin 64%, tapioca dextrin 75% and maltodextrin 8.5% (King *et al.*, 1976). Special modified starch (resulting from the reaction of starch with a lipophilic group) was very effective in encapsulating citrus and vegetable oils. It was as good an emulsion producer as gum arabic. If citrus oil and fruit juice solids are encapsulated by combining maltose and maltodextrin with gum arabic, and/or modified starch, a dense product is formed, having a bulk density of $0.5\,g\,cm^{-3}$, a moisture content of 2–8%, and less than 20% voids (Boskovic *et al.*, 1992).

Flavor composition and its careful creation can be directed to enhancing and modifying food taste (Subramaniam, 1996). An interesting example is the fixation of a flavor oil within a low cariogenic matrix of hydrogenated starch hydrolysate and maltodextrin with a DE of less than 20. The resultant mixture is then spray-dried to form a particulate product of $\sim$0.15–0.85 $g\,cm^{-3}$ bulk-free density (Subramaniam, 1996). Another manuscript describes flavor oil encapsulation in particulate polyol material. The formulation included poly-, mono- or disaccharide, water and flavor oil to form a preparation, with a T_g below room temperature. The extruded homogenate formed a solidified, hard, rubbery, amorphous material after cooling, containing the oil in its matrix (Blake & Attwool, 1996). Steam-distilled essential oils of basil, garlic and ginger were encapsulated in double-walled micro-capsules, built of maltodextrin and soy meal or maltodextrin and gum arabic, resulting in slow dissolution rate and good stability of the encapsulate (Sheen & Tasi, 1991). In addition to spray-drying, freeze-drying can be used as part of the encapsulation procedure (Maier *et al.*, 1987). Many other compounds can be added to the emulsion or formulation. *n*-Octyl alcohol, as a defoamer, and glutaraldehyde, as a cross-linking agent to render a coacervate of gum arabic and gelatin insoluble, may be used under the US Federal Food, Drug and Cosmetic Act in formulations of micro-capsules for encapsulating flavoring oils (Anon, 1968). Micro-capsules may be used under the US Federal Food, Drug and Cosmetic Act for encapsulating lemon, distilled lime, orange, peppermint and spearmint oils for use only in dry mixes for puddings and gelatin desserts. The capsules may be formulated from the following components: succinylated gelatin (max. 15% of the weight of the oil and micro-capsule; succinic acid content of the gelatin 4.5–5.5%) and arabinogalactan (Anon, 1968).

Spray-dry micro-encapsulation has been used to improve the stability of carotenoids in carrot pulp and paprika oleoresin. During the maturation of many fruits there is a change in color from green to orange or red. This is due to a loss of chlorophyll and the unmasking and synthesis of carotenoids. Carotenoids are C_{40}-isoprenoid compounds composed of isoprene units joined head-to-tail to form a system of conjugated double bonds. They are classified into two groups (Fig. 5.4), carotenes and xanthophylls. Examples of carotenes are carotenes α and β in carrots and lycopene in tomatoes; xanthophylls include capsanthin and capsorubin found in red pepper (Eskin Michael, 1990). The growing commercial interest in carotenoids as a food supplement has encouraged studies investigating the effects of processing conditions (inlet and outlet temperature and feed solids) on β-carotene content of spray-dried *Dunaliella salina* biomass. The lowest outlet temperature yielded higher carotenoid

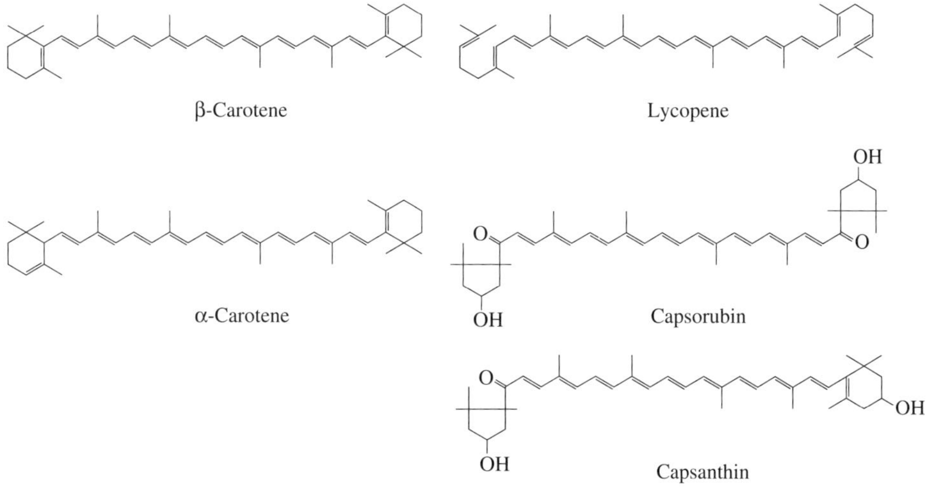

Fig. 5.4 Carotenes and xanthophylls.

recoveries (57–91%), while micro-encapsulation biomass gave 100% recovery (Leach *et al.*, 1998).

5.8.2 Flavoring of tea (dry vegetable matter) and other spices

Micro-encapsulated flavors (e.g. orange oil, cocoa flavor, etc.) mixed with a very small quantity of adhesive solution can be used to flavor dry vegetable matter, e.g. leaves and stems of tea, herbs, spices and vegetables. The adhesion solution contains water (in a quantity equal to or less than the amount absorbable by the vegetable material), gum arabic and sucrose. The dry vegetable material may be treated with water-miscible solvent prior to the adhesive solution. A long-standing example of the use of flavorings in this way is the treatment of *Earl Grey* tea with bergamot oil (Koene *et al.*, 1987).

Spices encapsulated in a carbohydrate matrix using pressure extrusion technology maintain free-flowing characteristics during long storage, and are stable to oxidation. The ingredients are soluble in hot or cold water, and are claimed to have a greater capacity for volatile retention than encapsulated spices produced by spray-drying (Andres, 1981).

5.8.3 Chewing gum

Chewing gum is a product consumed by all age groups. Since it stays within the mouth cavity for periods of a few minutes to hours (depending on the consumer), it is in the manufacturer's interest to prolong the chewing gum's flavor and juiciness-release capabilities for longer times to suit consumer demand. A flavor release composition (e.g. an extrudate or a particle) comprised a flavoring agent at 25% by weight, retained on non-thermoplastic silica, which was bound to a thermoplastic cellulosic material such as hydroxypropylmethyl cellulose (HPMC) (Song & Courtright, 1992). Another micro-encapsulated flavoring formulation for chewing gum proposed building a core that contains 20–80% flavoring agent, the rest being resin. The coating of the micro-capsule contained 45–49% gelatin, 45–49% gum arabic and 2–10% glutaraldehyde (Cherukuri *et al.*, 1993).

5.8.4 Spray-dried juices

Juices are a popular source of vitamins, sugars and carotenoids, among many other ingredients. They are sold as is (single strength), as concentrates, as blends with other juices or fruit concentrates, as part of squashes and after spray-drying for reconstitution purposes in drinks and other juice-containing products. Powdered juice is prepared from citrus, apple and other popular sources, but can also be produced from less common sources, such as jujube concentrate. The concentrated extract (26°Bx) is combined with a carrier (maltose) solution to reach 30°Bx and then spray-dried. Factors influencing the sensory evaluation and color of juice prepared from extract include carrier material composition and enzyme thermal treatment (Soon-An *et al.*, 1997).

5.8.5 Micro-encapsulated food ingredients

Micro-encapsulation processes are aimed at producing available compounds for use in the food and other industries. These include flavorings and other ingredients, which should be used as free-flowing powders and then released on a controlled basis during the food manufacture's process under appropriate conditions (Sinton, 1998). TasteTech's CR100 process uses carrier materials that, according to its claims, are tailored to suit the final foodstuff, into which the micro-encapsulated flavor is added. The process does not require high temperatures, and there is therefore less loss of flavor during the micro-encapsulation (Sinton, 1998). Spray-dried spinach juice was produced to obtain free-flowing powder by taking the vegetative tissue, and without blanching it, extracting the juice by twin-screw extractor, filtering it through a screen, adding 3% gum arabic as the carrier and spray-drying (Chiu & Wu-J, 1998). Carrageenan and maltodextrin have been used to produce micro-capsules of capsicum oleoresin, by homogenization, emulsification and spray-drying. The product contained 92.6% capsicum oleoresin (obtained by solvent extraction of the dried ripe fruit of *Capsicum annuum* with subsequent removal of the solvent, a clear-red, light-amber or dark-red viscid liquid with a characteristic odor and extremely strong bite), with a recovery rate of 91.5% (Xiang *et al.*, 1997). Successful encapsulation of blueberry flavor by gum arabic or maltodextrin has been reported (Li & Reineccius, 1995) for inclusion in pancake batters. Thermally controlled release properties were achieved with a second coating of gum arabic powder. Coating with fat of the same flavor improved flavor retention due to the fat's delayed release (Li & Reineccius, 1995). In general, spray-dried flavorings contribute greatly to bakery applications. During dough preparation or baking, flavoring is released and the carrier components decrease the volatilization rate (Dietrich, 1983).

5.8.6 Performance of different gums in encapsulation

Carrier ingredients can be taken from a long list of water-soluble polymers. Gum arabic serves as a very important ingredient in the encapsulation industry; thus it is not surprising that the performances of many other gums are compared to its role in similar compositions. Mesquite-bark exudate gum and gum arabic are similar in their levels of tannins and proteins, as well as in terms of their immunological characteristics (Goycoolea *et al.*, 1997). Both were tested in the spray-drying of orange peel essential oil. Gum arabic was superior in terms of yields of essential oil recovery (~100 versus 91% recovery with mesquite gum), but the mesquite gum proved its suitability and potential for replacing the gum arabic (Goycoolea *et al.*, 1997). Another report deals with both gum mixtures for oil encapsulation. The highest rate of encapsulation, ~94%, was observed when the gum arabic content was one and a half times that of the mesquite gum in the blend. The flavor intensity of orange essential oil encapsulated in gum arabic or a blend of gum arabic/mesquite gum capsules was the same (Beristain & Vernon-Carter, 1995). Other potential substitutes for gum arabic are starches of corn and amaranth (Chattopadhyay *et al.*, 1997). Other attempts to find substitutes for traditional gum arabic have involved the formation of chitinous material into capsules to serve as controlled-release vectors in food processing (Chen *et al.*, 1996). Other manuscripts describe the potential use of cyclodextrins (molecular capsules produced by enzymatic conversion of starch) as flavor carriers, for vitamins and cured meat pigment encapsulation (Allegre & Deratani, 1994). Gum arabic is used to a great extent by the food industry; it was therefore just a matter of time before new gum arabics that do not require manual tapping (removing sections of the bark with an axe, taking care not to damage the tree) would be tried in order to lower costs (LaBell, 1993). A similar idea involves developing special starches with functional properties similar to gum arabic. These show exceptional fat-binding and emulsion-stabilizing characteristics (Pegg & Shahidi, 1999).

5.8.7 Lipid encapsulation

Linoleic acid is an essential fatty acid that contains two double bonds. When it is present in food, the body can use it to synthesize fatty acids with three and four double bonds, as needed. In oxidative rancidity, oxygen is added to the double bond, causing a fatty acid to form such chemical compounds as peroxides, ketones and aldehydes, which produce noxious odors and flavors in fat. The oxidation rate of methyl linoleate can be decreased and controlled by using encapsulation with gum arabic via hot air or freeze-drying. Oxidation depends on the relative humidity during storage. Encapsulation by freeze-drying contributes to slower oxidation when compared to hot-air drying (Minemoto *et al.*, 1997). The two drying methods yield different morphologies, hinting at a great difference in the state of the encapsulated lipid within the matrices (Minemoto *et al.*, 1997). Fractured beads of maltodextrin produced by spray-drying were also reported to absorb three times their weight in hydrophobic oils, e.g. vegetable and fruit oils, and absorb smaller but still substantial amounts of hydrophilic oils (Anon, 1984). These beads are presented as carriers that do not agglomerate in oil. Mixtures of gum arabic and maltodextrin resulted in stronger beads that retain their properties in formulas containing crystalline hard structures (Anon, 1984). Another report discusses a food flavoring composition comprised of a complex of pyrolyzed fat/oil flavoring with a gelatinized amylose, or a blend of the complex with a protein hydrolysate (Chen Hsiung, 1998).

5.8.8 Flavoring for microwaveable foods

Attractions of microwave energy include the high rate of heating and absence of surface changes in the food. The greater convenience of microwaveable foods for consumers has led to substantial developments in products and packaging suitable for use in domestic microwave ovens (Fellows, 1990). Because microwave involves a specialized mode of operation, special flavorings need to be manufactured for heated, cooked, pasteurized and sterilized foods. Flavorings can be encapsulated by film coating, coacervation, extrusion or chill-spraying (Mancini, 1994). Successful carriers produced with acacia gum, modified starches, maltodextrins or based on other sources, provide optimum flavor release during microwave operation. Special attention should be paid to the fact that changes in various foods during microwave heating are highly variable (Mancini, 1994). Another manuscript describes flavorings for savory foods consisting of combined sautéed fat and spice flavor notes. These are specifically formulated for microwaveable and low-fat foods. Cooked fat notes are paired with spice notes to fit the flavor profiles of popular ethnic cuisines. Flavors come in both liquid and dry forms. Carriers for the spray-dried flavors include maltodextrin, acacia gum and non-fat dried milk (LaBell, 1992).

5.8.9 Flavors for beverages, soft drinks and sports drinks

Beverages are liquid foods consumed by drinking. Many liquids serve as beverages, including juices from fruits and vegetables served alone or in cocktails and punches; milk served alone or combined with other ingredients in chocolate drinks, milk-shakes, eggnogs, and sodas; carbonated beverages served alone or in mixtures with other beverages; coffee and tea (Gates, 1981). Gum arabic is used as an emulsifying agent for soft drink flavors and as an encapsulating agent for flavor oils (Wollen, 1983). Controlled-release formulations of electrolytes for sports persons performance, and other similar drinks are generally based on carbohydrates, biodegradable polymers, polyethylene glycols and polyvinyl alcohols (PVAs) (Booth, 1998).

A process for encapsulating aroma or flavor compounds for beverages by forming an oil-in-water emulsion from a hydrolyzed vegetable oil, an aqueous medium and a water-soluble, carbohydrate-based film-forming agent has been described (Liu & Rushmore, 1996). A sufficient amount of film-forming agent is added for the aqueous phase of the emulsion to contain at least 50% by weight of the soluble carbohydrate solids. The oil-in-water emulsion is sprayed onto a soluble beverage powder, whereupon the aqueous layer of each droplet evaporates to form the capsules. The moisture content of the soluble beverage powder after spraying is less than 5% by weight. The soluble beverage powder is then dissolved in hot water to release the flavor aroma (Liu & Rushmore, 1996). A powdered composition for the preparation of slow-CO_2-release effervescent soft drinks comprises an acid component and a carbonate component, each encapsulated in xanthan gum with or without the addition of sucrose or sucrose derivatives (Lavie, 1987). Another manuscript describes a procedure in which gellan beads containing beverage or food components such as flavorings, yeast or CO_2 are later introduced into the food for its enrichment (Chalupa, 1995).

5.8.10 Flavors for baking

Baking is a unit operation that uses heated air to alter the eating quality of foods. Baking is usually applied to flour-based products. A secondary purpose of baking is preservation via the destruction of microorganisms and reduction in the water activity at the surface of the food (Fellows, 1990). Aromas are produced during baking as a result of Maillard Browning reactions between sugars and amino acids. Other reactions involve sugar caramelization, and fatty acid oxidation to aldehydes, lactones, ketones, alcohols and esters. These processes are also influenced by the moisture content within the surface layers (van Quaquebeke, 1976). Flavor preparation methods for inclusion in bakery products include mixing the flavor with the carrier material (gelatin, gelatin–gum arabic complexes and cellulose derivatives) followed by spray-drying. The resultant material is later easily mixed with the baking ingredients. Its effectiveness as a flavoring material should be tested by sensory methods (van Quaquebeke, 1976). The capsules used in such cases have sizes of 5–2000 μm, and can serve to introduce more exotic aromas, such as ginger, to bakery products (Richards, 1971).

5.8.11 Encapsulated salts

The taste of reduced-salt foods and beverages is improved by incorporating edible encapsulated ammonium salts. These salts can be prepared from ammonia recovered during spray-drying of fermented soy sauce or from acid hydrolysis of a protein. Carrier agents include maltodextrin, gum arabic and gelatinized starches, particularly high-amylopectin hydrolysates debranched at their 1,6-α-D-glycosidic linkages (Lee & Tandy, 1994).

5.9 Interactions between volatile compounds and hydrocolloids

During the production of aroma compounds, interactions between volatile compounds and hydrocolloids can occur. The sensory evaluation of flavored solutions with tastes of mushroom (1-octen-3-ol), garlic (dialyl disulfide/diallyl sulfide) and butter (diacetyl) demonstrated that in-mouth sensations were the greatest with water, intermediate with 0.1% xanthan and the lowest with 0.3% guar gum. This is a result of their viscosity at mouth shear rates. Addition of xanthan and guar gum lowered flavor release. Weak and reversible hydrogen bonding between xanthan and 1-octen-3-ol has been demonstrated by exclusion chromatography (Yven *et al.*, 1998).

5.10 Micro-capsule micro-structure

The slow-release properties of micro-capsules are dependent on their micro-structure. This, in turn, is based on the processing parameters and can be investigated by techniques developed for scanning electron microscopy (SEM) sample preparations. These techniques include the embedding of a specimen in an apolar resin and partial polymerization, followed by observation of the inner and outer structure of the examined capsule. The core material is organized in the solid wall matrix. With spray-dried gum arabic micro-capsules one or more internal voids were observed. Such micro-capsules had indentations and caps on the exterior of the capsules, and these in turn were influenced by the total solids concentration in the sprayed emulsion and working parameters (Rosenberg *et al.*, 1988). Starch granules and bonding agents or water-soluble polymers can be spray-dried to obtain porous spheres. These structures can carry a variety of food ingredients (flavors, essences, etc.) for future slow release. Special uses of such structures are in dry mixes for baking or reconstitution, or for prolonged release in chewing gum (Zhao & Whistler, 1994). Special micro-capsule structures can be achieved by changing the drying procedure from spray to cold-dehydration. In this method, an emulsion of the core material in a gum solution is injected into ethanol and the micro-capsule slurry is then dried in a vacuum oven (Zilberboim *et al.*, 1986). Enhanced textural properties can be achieved by using amylase-treated granular starches as a micro-porous matrix carrier for functional compositions, which are released under the influence of mechanical compression, via diffusion into a surrounding fluid, or as a result of degradation of the matrix. The micro-porous starch granules are chemically derivatized, to enhance the absorptive and structural properties (Whistler, 1991).

5.11 References

Allegre, M. & A. Deratani (1994) Cyclodextrin uses: from concept to industrial reality. *Agro Food Industry Hi Tech.*, **5**(1), 9–17.

Anandaraman, S. & G.A. Reineccius (1986) Stability of spray-dried orange peel oil. *Food Technol.*, **40**(11), 88.

Anderson, D.M.W., D.M. Brown Douglas, N.A. Morrison *et al.* (1990) Specifications for gum arabic (*Acacia senegal*): analytical data for samples collected between 1904 and 1989. *Food Addit. Contam.*, **7**(3), 303–321.

Andres, C. (1981) Microencapsulation protects flavor, aroma volatiles until released in water. *Food Processing*, **42**(10), 38.

Anon (1968) Food additives. Microcapsules for flavoring oils. Federal Register, 33 (242, Dec. 13), 18488 & Federal Register, 33 (232, Nov. 28) 17752.

Anon (1984) Flavors and oils carried on maltodextrin 'buds'. *Food Eng.*, **56**(11), 78.

Anon (1994) Encapsulation: a flavor survival strategy. *Prepared Foods*, **163**(3), 73.

Baisier, W. & G.A. Reineccius (1989) Spray drying of food flavors: factors influencing shelf-life of encapsulated orange peel oil. *Perfum. Flav.*, **14**, 48–53.

Balke, W.J. (1984) The use of natural hydrocolloids as thickeners, stabilizers and emulsifiers. Special report, New York State Agricultural Experiment Station, No. 53, 31–34.

Bangs, W.E. & G.A. Reineccius (1990) Characterization of selected materials for lemon oil encapsulation by spray drying. *J. Food Sci.*, **55**(5), 1356–1358.

Barnes, J.M. & J.A. Steinke (1987) Encapsulating matrix composition and encapsulate containing same. United States Patent 4,689,235.

Beck, E.E. (1972) Essential oils. United States Patent 3,704,137.

Benoit, J.P., S. Morteau & J. Richard (1997) Microparticles of crosslinked polymer containing an organic substance, and process for manufacture of these microparticles. French Patent Application FR 2,748,673A1.

Beristain, C.I. & E.J. Vernon-Carter (1995) Studies on the interaction of arabic (*Acacia senegal*) and mesquite (*Prosopis juliflora*) gum as emulsion stabilizing agents for spray-dried encapsulated orange peel oil. *Drying Technol.*, **13**(1/2), 455–461.

Bhandari, B.R., E.D. Dumoulin, H.M.J. Richard, I. Noleau & A.M. Lebert (1992) Flavor encapsulation by spray drying: applications to citral and linalyl acetate. *J. Food Sci.*, **57**(1), 217–221.

Bhandari, B.R., B.R. D'Arcy & L.L.T. Bich (1998) Lemon oil to β-cyclodextrin ratio effect on the inclusion efficiency of β-cyclodextrin and the retention of oil volatiles in the complex. *J. Agric. Food Chem.*, **46**(4), 1494–1499.

Bhandari, B.R., E.D. Dumoulin, H.M.J. Richard, I. Noleau & A.M. Lebert (1992) Flavor encapsulation by spray drying: application to citral and linalyl acetate. *J. Food Sci.*, **57**, 217–230.

Blake, A. & P. Attwool (1996) Particulate flavor compositions and process to prepare same. PCT International Patent Application WO 96/11589 A1.

Booth, G.P. (1998) Controlled-release formula for performance drinks. PCT International Patent Application WO98/20755 A1.

Boskovic, M.A., S.M. Vidal & F.Z. Saleeb (1992) Spray-dried fixed flavorants in a carbohydrate substrate and process. United States Patent 5,124,162.

Chalupa, W.F. (1995) Gellan gum flavor beads. United States Patent 5,456,937.

Chattopadhyay, S., R.S. Singhal & P.R. Kulkarni (1997) Optimization of conditions of synthesis of oxidized starch from corn and amaranth for use in film-forming applications. *Carbohydrate Polymers*, **34**(4), 203–212.

Chen, R.H., M.L. Tsaih & W.C. Lin (1996) Effects of chain flexibility of chitosan molecules on the preparation, physical and release characteristics of the prepared capsule. *Carbohydrate Polymers*, **31**(3), 141–148.

Chen Hsiung, L.E. (1998) Flavor composition. United States Patent 5,780,089.

Cherukuri, S.R., K.P. Raman & G. Mansukhani (1993) Microencapsulated flavoring agents and methods for preparing same. United States Patent 5,266,335.

Chiu, R.H. & S.B. Wu-J (1998) Processing of spinach powder and spinach flakes. *Food Sci., Taiwan*, **25**(4), 446–455.

Connolly, S., T.C. Fenyo & M.C. Vandevelde (1987) Heterogeneity and homogeneity of an arabinogalactan-protein-Acacia senegal gum. *Food Hydrocolloids*, **1**(5/6), 477–480.

Dietrich, H. (1983) Importance of flavorings in bakeries. *Brot & Backwaren*, **31**(5), 115,118.

Eskin Michael, N.A. (1990) In: *Biochemistry of Foods*. Academic Press, San Diego, p. 102.

Fellows, P.J. (1990) In: *Food Processing Technology*. Ellis Horwood Limited, New York, pp. 312 & 347.

Gates, J.G. (1981) *Basic Foods*. Holt, Rinehart and Winston, NY.

Goldner, W.R. (1996) Flavor emulsions and beverages containing leucaena gum. United States Patent 5,508,059.

Goubet, I., J.L. Le Quere & A.J. Voilley (1998) Retention of aroma compounds by carbohydrates: Influence of their physicochemical characteristics and their physical stare: A review. *J. Agric. Food Chem.*, **46**, 1981–1990.

Goycoolea, F.M., A.M.C. de-la-Barac, J.R. Balderrama & J.R. Valenzuela (1997) Immunological and functional properties of the exudate gum from northwest Mexican mesquite (*Prosopis* spp.) in comparison with gum arabic. *International J. Biol. Macromolecules*, **21**(1/2), 29–36.

Harrison, M. & B.P. Hills (1996) A mathematical model to describe flavor release from gelatin gels. *International J. Food Sci. Technol.*, **31**(2), 167–176.

Hassan, H.M., A.A. Sayed & C.J. Mumford (1996) Volatiles retention in the drying of skin forming materials. III. Heat sensitive materials. *Drying Technol.*, **14**(3/4), 581–593.

Heinrich, H.W., B. Kuklinski, U. Meyer, F. Westphal & J. Teller (1998) Flavor neutral microcapsules, process for their manufacture, and their uses. German Federal Republic Patent Application DE 19644343A1.

Imeson, A.P. (1992) Exudate gums. In: (Imeson, A.P., Ed.), *Thickening and Gelling Agents for Food*. Blackie Academic & Professional, Bishopbriggs, Glasgow, pp. 66–97.

Kayaert, G., P. Tobback, E. Maes, J. Flink & M. Karel (1975) Retention of volatile organic compounds in a complex freeze-dried food gel. *J. Food Technol.*, **10**(1), 11–18.

Kerkhof, P.J.A.M. & H.A.C. Thijssen (1974) Retention of aroma components in extractive drying of aqueous carbohydrate solutions. *J. Food Technol.*, **9**(4), 415–423.

Kim, Y.D. & C.V. Morr (1996) Microencapsulation properties of gum arabic and several food proteins: spray-dried orange oil emulsion particles. *J. Agric. Food Chem.*, **44**, 1314–1320.

King, A.H. (1995) Encapsulation of food ingredients. A review of available technology, focusing on hydrocolloids. *ACS Symposium Series*, **590**, 26–39.

King, W., P. Trubiang & P. Perry (1976) Modified starch encapsulating agents offer superior emulsification, film forming, and low surface oil. *Food Product Development*, **10**(10), 54, 56–57.

Koene, C.H., C. Vos & J. Brasser (1987) Process for flavoring dry vegetable matter. European Patent EP 0,109,698 B1.

Kollengode, A.N. & M.A. Hanna (1996) Influence of storage temperature on flavor retention of internally flavored extrudates. *IFT Annual Meeting: Book of Abstracts*, p. 111.

LaBell, F. (1992) Sauteed flavors and savory, fried taste. *Food Processing, USA*, **53**(11), 28, 33.

LaBell, F. (1993) Gum arabic effective for dry flavors. *Food Processing, USA*, **54**(5), 82–83.

Lavie, L. (1987) Water soluble non-hygroscopic powdered composition for preparation of drinks with prolonged gas release and process for its preparation. European Patent Application EP 0,233,839 A1.

Leach, G., G. Oliveira & R. Morais (1998) Spray-drying of *Dunaliella salina* to produce a β-carotene rich powder. *J. Industrial Microbiol. Biotechnol.*, **20**(2), 82–85.

Lee, E.C. & J.S. Tandy (1994) Taste-enhancement of sodium chloride reduced compositions. United States Patent 5,370,882.

Lencki, R.W.J., R.J. Neufeld & T. Spinney (1989) Method of producing microspheres. United States Patent US 4,822,534.

Levine, H. & L. Slade (1989) Interpreting the behavior of low moisture foods. In: (Hardmann, T.M., Ed.), *Fundamental Aspects of the Dehydration of Foodstuffs*. Elsevier Applied Science, NY, pp. 71–134.

Li, H.C. & G.A. Reineccius (1995) Protection of artificial blueberry flavor in microwave frozen pancakes by spray drying and secondary fat coating processes. *ACS Symposium Series*, **590**, 180–186.

Liu, R.T. & D.F. Rushmore (1996) Process for making encapsulated sensory agents. United States Patent 5,580,593.

Maier, H.G., K. Moritz & U. Ruemmler (1987) Thermostable binding of aroma compounds to starch. I. Binding by freeze-drying. *Starch*, **39**(4), 126–131 (in German).

Mancini, L. (1994) Delivering flavor into the 21st century. *Food Eng.*, **66**(8), 104–105.

Miller, D.H. & J.R. Mutka (1985) Preparation of solid essential oil flavor composition. United States Patent 4,499,122.

Miller, D.H. & J.R. Mutka (1986) Solid essential oil composition. United States Patent 4,610,890.

Minemoto, Y., S. Adachi & R. Matsuno (1997) Comparison of oxidation of methyl linoleate encapsulated with gum arabic by hot air drying and freeze drying. *J. Agric. Food Chem.*, **45**(12), 4530–4534.

Mothes, H. (1997) Encapsulated flavors for sophisticated products. *European Food & Drink Review*, Autumn, 20–21.

Nelson, K. & T.P. Labuza (1992) Understanding the relationships between water and lipid oxidation rates using water activity and glass transition theory. In: (St. Angelo, A., Ed.), *Lipid Oxidation in Foods*. American Chemical Society, Washington D.C., pp. 91–101.

Nussinovitch, A. (1997) In: *Hydrocolloid Applications, Gum Technology in the Food and Other Industries*. Blackie Academic & Professional, UK, pp. 125–130 & pp. 247–250.

Pegg, R.B. & F. Shahidi (1999) Encapsulation and controlled release in food preservation. In: (Shafiur Rahman, M., Ed.), *Handbook of Food Preservation*. Marcel Dekker, Inc., New York and Basel, pp. 611–667.

Phillips, G.O. & P.A. Williams (2000) Handbook of Hydrocolloids. CRC, Woodhead Publishing Limited, Cambridge, England, pp. 155–169.

Pothakamury, U.R. & G.V. Barbosa Canovas (1995) Fundamental aspects of controlled release in foods. *Trends Food Sci. Technol.*, **6** (12), 397–406.

Ré, M.I. (1998) Microencapsulation by spray drying. *Drying Technol.*, **16**(6), 1195–1236.

Reineccius, G.A. (1989) Flavor encapsulation. *Food Rev. Intern.*, **5**, 147.

Reineccius, G.A. (1991) Carbohydrates for flavor encapsulation. *Food.Technol.*, 144–149.

Reineccius, G.A. & W.E. Bangs (1985) Spray drying of food flavors. III. Optimum infeed concentrations for the retention of artificial flavors. *Perfum. & Flav.*, **10**(1), 27–29.

Reineccius, G.A., F.M. Ward, C. Whorton & S.A. Andon (1995) Developments in gum acacia for the encapsulation of flavors. *ACS Symposium Series*, **590**, 161–168.

Richards, P.A. (1971) Microencapsulation. Food Manufacture; March April, Ingredients and raw materials Survey, 13.

Risch, S.J. (1986) Encapsulation of flavors by extrusion. In: (Risch, S.J. & G.A. Reineccius, Eds.), *Flavor Encapsulation*. American Chemical Society, Washington, DC, pp. 103–109.

Risch, S.J. (1995) Review of patents for encapsulation and controlled release of food ingredients. *ACS Symposium Series*, **590**, 196–203.

Rosenberg, M. & I.J. Kopelman (1983) Microencapsulation of food ingredients-processes, application and potential. Proceedings of the 6th International Congress of Food Science and Technology, **2**, 142–143.

Rosenberg, M., Y. Talmon & I.J. Kopelman (1988) The microstructure of spray-dried microcapsules. *Food Microstructure*, **7**(1), 15–23.

Sayed, A.A., H.M. Hassan & C.J. Mumford (1996) Volatiles retention in the drying of skin forming materials. I. Materials which gelatinize at moderately high temperature. *Drying Technol.*, **14**(3/4), 529–563.

Sheen, L.Y. & S.J. Tasi (1991) Studies on spray-dried microcapsules of ginger, basil and garlic essential oils. *J. Chinese Agric. Chem. Soc.*, **29**(2), 226–237.

Sinton, R. (1998) TasteTech-the taste of things to come. *Nutr. Food Sci.*, **4**, 227–230.

Song, J.H. & S.B. Courtright (1992) Flavor releasing structures for chewing gum. United States Patent 5,128,155.

Soon-An, D., K.L. Woo & D.S. Lee (1997) Processing of powdered jujube juice by spray drying. *J. Korean Soc. Food Sci. Nutr.*, **26**(1), 81–86.

Subramaniam, A. (1996) Particulate hydrogenated starch hydrolysate based flavoring materials and use of same. United States Patent 5,506,353.

Swisher, H.E. (1957a) Solid essential oil-containing components. United States Patent 2,809,895.

Swisher, H.E. (1957b) Solid essential oil flavoring components. United States Patent 3,041,180.

Taylor, A.H. (1983) Encapsulation systems and their applications in the flavor industry. *Food Flavoring Ingredients Packaging and Processing*, **5**(9), 48–49, 51–52.

Thevenet, F. (1986) Acacia gums: stabilizers for flavor encapsulation. In: (Risch, S.J. & G.A. Reineccius, Eds.), *Flavor Encapsulation*. American Chemical Society, Washington, DC, pp. 37–44.

Trubiano, P.C. (1995) The role of specialty food starches in flavor encapsulation. *ACS Symposium Series, Flavor Technology Physical Chemistry Modification and Process*. Washington, DC, pp. 244–253.

Trubiano, P.C. & L. Lacourse (1986) Emulsion-stabilizing starches: use in flavor encapsulation. In: (Risch, S.J. & G.A. Reineccius, Eds.), *Flavor Encapsulation*. American Chemical Society, Washington, DC, pp. 45–54.

Tuley, L. (1996) Breaking free. Update on controlled flavor release. *International Food Ingredients*, **3**, 14–16, 18.

van Quaquebeke, H. (1976) Essences: their applications and characterization in bakery products. *Getreide,-Mehl-und-Brot*, **30**(2), 36–40.

Veronese, J. (1974) Microencapsulation of aromas, odors, active substances and oils. *Seifen Oele Fette Wachse*, **100**(12), 281–283, 293–294.

Whistler, R.L. (1991) Microporous granular starch matrix compositions. United States Patent 4,985,082.

Wollen, A. (1983) Spray dried gum aids soft drinks flavoring. *Soft Drinks Trade J.*, **37**(3), 110–111.

Xiang, Y.F., D.Y. Liu, J.Z. Yang, P. Jin & M. Cheng (1997) Microencapsulation of capsicum oleoresin. *Food Sci., China*, **18**(11), 27–30.

Young, D.K. (1996) Encapsulation properties of several food proteins. *Dissertation Abstracts International B*, **56**(9), 4661.

Yven, C., E. Guichard, A. Giboreau & D.D. Roberts (1998) Assessment of interactions between hydrocolloids and flavor compounds by sensory, headspace and binding methodologies. *J. Agric. Food Chem.*, **46**(4), 1510–1514.

Zhao, J. & R.L. Whistler (1994) Spherical aggregates of starch granules as flavor carriers. *Food Technology*, **48**(7), 104–105.

Zilberboim, R., I.J. Kopelman & Y. Talmon (1986) Microencapsulation by dehydration liquid: a microstructural study by scanning electron microscopy. *J. Food Sci.*, **51**(5), 1307–1310.

Chapter 6

Immobilization for Food and Biotechnological Purposes

6.1 Definition, aims and features of immobilized preparations

Immobilization relates to the prevention of free cell movement by natural or artificial means (Bucke, 1983). At one extreme, the immobilized cells can be fully capable of division; at the other, they may possess only one type of enzymatic activity (Mattiasson, 1983). Support materials for immobilization are divided into two types: organic and inorganic. The discussion in this chapter is limited to organic support materials, which fall into three categories: polysaccharides, proteins and synthetic polymers. Polysaccharides include agar, agarose, alginate, carrageenan, cellulose, dextran and pectate, among others; proteins include collagen, egg white and gelatin, and the synthetic polymers include polyacrylamide, phenolic resins, polystyrene and polyurethane (Nussinovitch, 1997).

In general, the desirable features for immobilized cell and enzyme preparations are: high biocatalytic activity, long-term stability of the biocatalyst, possibility of regenerating the biocatalyst, low loss of activity during immobilization, low leakage of cells, non-compressible particles, high resistance to abrasion, resistance to microbial degradation, low diffusional limitations, spherical shape, high surface area, appropriate density for the reactor type, technique simplicity, inexpensive support materials, and non-toxicity of those materials (Linko, 1974; Nussinovitch, 1997; Olston & Korus, 1976; Smiley, 1976; Smiley & Strandberg, 1972; Tampion & Tampion, 1987). Cell immobilization offers advantages for the food-processing industry, including enhanced fermentation productivity, cell stability and reduced downstream processing costs due to facilitated cell recovery and recycling. A summary of the various immobilization methodologies, including adsorption, entrapment, covalent binding and micro-encapsulation, can be found elsewhere (Groboillot *et al.*, 1993). Examples of interest to the food industry are provided, together with a review of the physiological effects of immobilization. Topics in process engineering include immobilized cell bioreactor configurations and the scale-up potential of the various immobilization techniques (Groboillot *et al.*, 1993).

6.2 Immobilization media and techniques

6.2.1 Immobilization by gelling agents originating from seaweed extracts

Cells are often entrapped within a gel matrix. A wide range of characteristics are attributed to gels as an entrapment medium. On the one hand, they include macromolecules held together by relatively weak inter-molecular forces, such as hydrogen-bonding or ionic cross-bonding by di- or multi-valent cations. On the other hand, strong covalent bonding, where the lattice in which the cells are entrapped is considered to be one vast macro-molecule, is limited only by the particle size in the immobilized cell preparation (Nussinovitch, 1997; Nussinovitch *et al.*, 1994; Tampion & Tampion, 1987).

The major categories of entrapment have been reviewed (Bucke, 1983; Cheetham, 1980; Mattiasson, 1983; Nussinovitch, 1997; Nussinovitch *et al.*, 1994). They include some commonly used, single-step entrapment methods, such as the simple gelation of macro-molecules by lowering or raising temperatures, using hydrocolloids such as agar, agarose, κ-carrageenan and chitosan, and proteins such as gelatin and egg whites, among others. However, these preparations generally suffer from low mechanical strength and possible heat damage. Another simple single-step entrapment method is the iontropic gelation of macro-molecules by di- and multi-valent cations, alginate (Fig. 6.1) and LMP being good examples of this. The limitations of such a system are: low mechanical strength and breakdown in the presence of chelating agents. Another overview of hydrogels for cell immobilization has been published (Jen *et al.*, 1996). Current developments in the immobilization of mammalian cells in hydrogels are also reviewed. Discussions cover hydrogel requirements for use in adhesion, matrix entrapment and micro-encapsulation, the respective processing methods, as well as current applications (Jen *et al.*, 1996).

Agar is a complex water-soluble polysaccharide. It is a hydrophilic colloid extracted from certain marine algae of the class Rhodiphyceae. Agar holds a unique

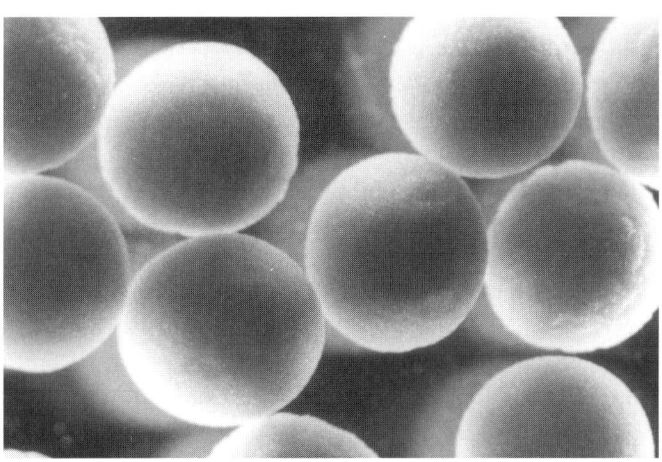

Fig. 6.1 Hydrocolloid beads of alginate (magnification × 15). (Courtesy of Dr. Y. Tal.)

position in its commercial importance because it forms firm gels at concentrations as low as 1% (Davidson, 1980; Nussinovitch, 1997). The term agar characterizes a family of polysaccharides with a backbone of alternate 1-3-linked and 1-4-linked D- and L-galactose residues (Davidson, 1980). Agarose, the gelling agent in agar, was used (Khachatourians *et al.*, 1982) to entrap whole cells of *Escherichia coli* and a-nucleate mini-cells produced by a mutant with defective cell division at 50°C. Agarose entrapment is one of the mildest methods available, allowing for the retention of cell viability. Special grades of agarose which have lower gelling temperatures are supplied by some manufacturers. One of these agarose grades was used (Brodelius & Nilsson, 1980) to immobilize *Catharanthus roseus* cells. These cells retained cellular integrity, as evidenced by their respiratory activity and susceptibility to plasmolysis, and were capable of growth. Many other applications, including the entrapment of algae and blue-green bacteria, have been reported (Wiksstrom *et al.*, 1982).

For general use, a dilute solution, usually 2–4% (w/v) agar, is prepared in a medium suitable for the particular cells being immobilized. The agar is dissolved at ~90°C before any cells are added to the solution; this addition is effected at a temperature a bit higher than the setting point to minimize heat damage. Usually, a 1:1 ratio of cell suspension to agar solution is used (Brodelius & Nilsson, 1980). The gel can be formed in sheets or slabs of any desired thickness, to be cut later into smaller pieces. Perforated molds are used to produce short cylindrical beads (Brodelius & Nilsson, 1980). Spherical beads can be obtained by dropping the hot solution of the preparation into ice-cold buffer. Another technique is to add the hydrocolloid solution to a heated oil bath to produce a hot emulsion that is later cooled to produce the gel beads (Wiksstrom *et al.*, 1982). Modifications of such preparations can be found in the literature (Banerjee *et al.*, 1982). Other reported entrapments include the immobilization of *Bacillus licheniformis* in agar and alginate gels (Dobreva *et al.*, 1996).

Carrageenans are water-soluble gums which occur in certain species of red seaweed. Chemically, they are sulfated linear polysaccharides of D-galactose and 3,6-anhydro-D-galactose. The use of κ-carrageenan as an immobilization medium was reported by Chibata (1981). It is more readily used than agar as a growth substrate by microorganisms. A *Saccharomyces cerevisiae*–carrageenan mixture was pumped for gelation into a 2% (w/v) potassium chloride solution (Wang & Hettwer, 1982). A large number of cells were found to be loaded within the beads. Ten times more cells were observed in a batch reactor holding the immobilized cells than with free cells. Although both followed the same typical growth curve, the immobilized cells reached the stationary-phase plateau at a higher cell density, taking about twice the time of the free yeast cells. Cell leakage could be reduced from carrageenan beads by increasing the potassium chloride concentration to 4% (w/v), but with a concomitant loss in cell viability (Wang *et al.*, 1982). Eighteen species (Mattiasson, 1983) were entrapped by such beads.

κ-Carrageenan gelation in the presence of potassium chloride was demonstrated experimentally by Krouwel *et al.* (1982). These authors claimed that carrageenan gels (beads) are superior to agar and inferior to calcium alginate. A comparison between κ-carrageenan and calcium alginate as entrapment media was reported. A small advantage of the κ-carrageenan over calcium alginate was reported (Grote *et al.*, 1980). Addition of LBG to κ-carrageenan can help in producing gel beads. Non-food applications of such beads exist, e.g. antibiotic production (Deo & Gaucher, 1983). Combined matrices of κ-carrageenan and calcium alginate were used for hormone transformation (Kim *et al.*, 1982). The κ-carrageenan matrix is not capable of providing full protection against heating processes (Brodelius & Nilsson, 1980). These examples of non-food applications are given here to emphasize the possibility of using their benefits in future food applications.

Immobilization of *Bacillus acidocaldarius* whole-cell *Rhodanese* in polysaccharide matrices of calcium alginate, κ-carrageenan and chitosan, and in an insoluble gelatin gel has been reported (Deriso *et al.*, 1996). The results obtained with the different immobilisates in terms of activity yield, possibility of regeneration and operative stability were evaluated with the aim of setting up a continuous system. This was achieved by entrapping *B. acidocaldarius* cells in an insolubilized gelation matrix. The latter, in the form of a thin membrane, was employed in a custom-conceived, plug-flow reactor (Deriso *et al.*, 1996).

Alginates are sold for specific applications. The contents of manuronic and guluronic residues and their ratio are not measured by the manufacturer. Instead they are estimated from data on the seaweed source and the blended quantities used to achieve the desired gel strength and solution viscosity. Sodium alginate is a commercially available product sold in bulk quantities; calcium, potassium and ammonium alginates, as well as alginic acid, are also available (Nussinovitch, 1997). It is important to note that alginic acid, the free acid form of alginate, has limited stability: to stabilize it and make it water-soluble, the acid is transformed into a commercial alginate by the incorporation of salts such as sodium carbonate, potassium carbonate, ammonium hydroxide, magnesium hydroxide, calcium chloride or propylene oxide (Nussinovitch, 1997). Many reports on the immobilization of enzymes, and microorganisms (yeasts, lactic acid bacteria and others), and on the immobilization of plant, fungal and animal tissues in alginates can be found (Jankovsky & Vasakova, 1996; Ogbonna *et al.*, 1989).

In fact, sodium alginate gelation by cross-linking via di- or multi-valent ions is the most studied method of entrapment (Fig. 6.2). Such gelation is easy to perform: cells are mixed with a sodium alginate solution and added dropwise into dilute baths of calcium chloride. The method retains cell viability, but cell decomposition can occur via the introduction of chelating agents such as phosphate and/or citrate buffers. The strength and porosity of the beads can be somewhat controlled by choosing an alginate with a specific composition. The higher the L-guluronic acid content, the stronger the gel. If alginate with a higher proportion of D-mannuronic acid is chosen for immobilization, beads with larger internal pore sizes are produced. As the concentration of alginate in the beads increases, so does the mechanical strength. An increase in the cell mass has the reverse effect. Diffusion coefficients of glucose and ethanol in alginic membranes were studied (Hannoun & Stephanopoulos, 1986). Diffusivity decreased with increasing concentrations of alginate. In a work describing alginate-immobilized *Chlorella* (Dainty *et al.*, 1986), the importance of providing suitable environmental conditions, especially if prolonged use of alginate is desired,

(a)

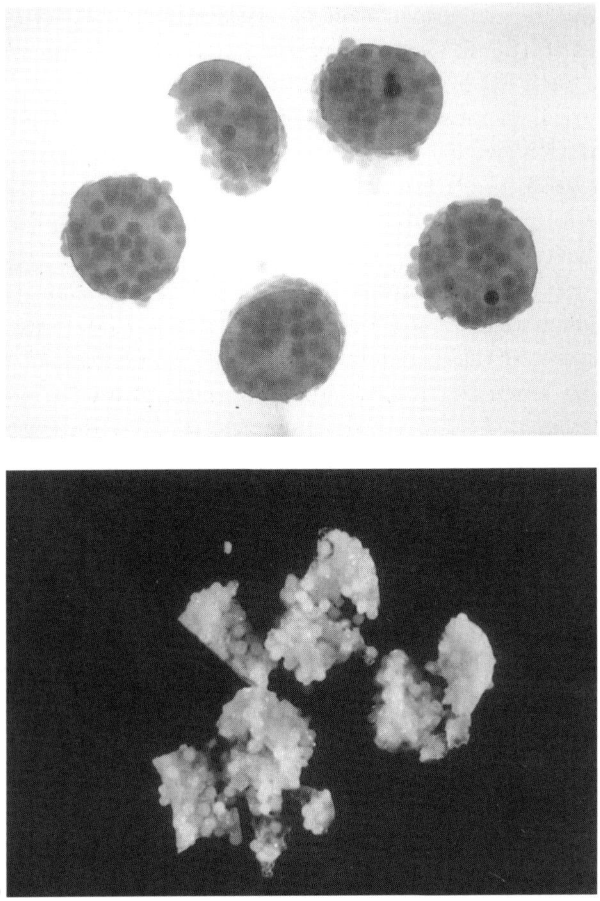

(b)

Fig. 6.2 (a) Immobilized beads within a gel; (b) ruptured–compressed cylindrical gel. A convenient way of producing a multi-textured food, composite material or complex entrapment.

was discussed. Trivalent ions were beneficial in achieving greater alginate bead strengths (Rochefort *et al.*, 1986).

A different approach for achieving predetermined properties of entrapment media consists of two-step methods. Combinations of different methods could provide means of overcoming evident shortcomings. Alginate-bead stabilization using a combination of three methods was discussed by Birnbaum *et al.* (1981). Firstly, alginate beads were sequentially treated with polyethyleneimine–HCl and glutaraldehyde, the former being allowed to infiltrate the beads for 24 h while the latter was applied as a 1% (v/v) solution at pH 7 for 1 min. The beads were then washed in water. In the second method the alginate was first activated for 1 h with a mixture of *N*-hydroxysuccinimide and 1-ethyl-3-(3-dimethyl-amino-propyl)-carbodiimide. Beads were formed by conventional addition to calcium chloride solution and allowed to cure for 1 h. The beads were then stabilized by a 24 h treatment in a defined suspension medium containing polyethyleneimine, and washed in water. In the third method the sodium alginate was pre-reacted with sodium meta-periodate for 1 h. The cells, in an equal quantity of unreacted sodium alginate, were then mixed in and beads formed. After curing for 1 h the beads were treated with polyethyleneimine–HCl and washed in water. The exact mechanism of the three methods is not fully understood, but they all probably involve some ionic or covalent binding of the cells themselves. Even an element of direct coupling of the cells to the activated alginate has been considered (Birnbaum *et al.*, 1981).

6.2.2 Immobilization by proteins

Proteins are widely used in the food business for many applications; therefore, their contribution to immobilization is interesting and challenging. They are inexpensive available sources for flexible preparations, which can be used in a variety of biotechnological applications. Collagen is an animal protein derived from connective tissues. To stabilize its structure, it is usually tanned using glutaraldehyde, which cross-links the protein molecules. Cell immobilization also makes use of this reaction for matrix stabilization. However excessive exposure to glutaraldehyde damages the cell function, and optimal conditions need to be chosen carefully before preparation. An interesting overview of the entrapment of eight types of bacteria within a membrane cast from tanned collagen can be found in Venkatasubramanian and Vieth (1979). However, the method described by those authors is less favored because gelatin produced from the same source is of higher quality. Most types of protein-cell immobilization make use of this reaction to stabilize the matrix, therefore it is of interest, especially if egg white, gelatin, gelatin–agarose and gelatin–alginate copolymers, all cross-linked by glutaraldehyde, are used (Brodelius & Nilsson, 1980).

6.2.3 Immobilization by synthetic gums

Other single-step entrapment methods include the use of synthetic polymers produced by chemical or photochemical reaction. Typically used materials include epoxy resins, polyacrylamide and polyurethane. In such cases, gel precursors are often toxic and some gels have low mechanical strength. Another single-step entrapment method involves precipitation from an immiscible solvent, as in the case of cellulose triacetate and polystyrene. Here the solvents are often toxic (Nussinovitch, 1997).

Polyacrylamide is a synthetic polymer which exhibits different properties depending on the reagents and conditions used during its preparation. An early study of cell immobilization by entrapment in polyacrylamide was reported by Mosbach and Mosbach (1966). The mechanical strength of the gel increased in proportion to the square root of the concentration of the acrylamide monomer used with a given fixed concentration of cross-linking agent. This was followed by a decrease in the gel's pore size which imposes some diffusional restrictions. The least cell damage is obtained using relatively high concentrations of tetramethylethylenediamine and ammonium persulfate. The mechanical properties of the polyacrylamide gels were compared to those of alginate, agar and κ-carrageenan (Krouwel *et al.*, 1982). Only the alginate gels showed a non-homogeneous gel structure, with a weaker inner portion and a stronger outer layer. Agar had very poor strain-resisting properties. The authors do not specify the exact conditions for polymerization and so the results for the polyacrylamide are difficult to interpret (Krouwel *et al.*, 1982). The achievement of maximal mechanical strength was also reported (Klein *et al.*, 1978). The diffusional properties of polyacrylamide gels containing immobilized *Alcaligenes faecalis* were studied (Wheatley & Phillips, 1983). These properties are clearly of paramount operational significance. In fact, the unsuitability of polyacrylamide gels for particular cell types was demonstrated. Positive results on the entrapment of *E. coli* in polyacrylamide were reported (Bang *et al.*, 1983). Polyacrylamide was found superior to polymethacrylamide or polyepoxide in terms of the production of tryptophan from serine and indole. Many other synthetic polymeric gel systems used for entrapment are mentioned in the literature. They include copoly-(styrene-maleic acid), and polyethylene glycol methacrylate, polyisocyanates and polyurethane, among many others. Polyacrylamide gels have been used to entrap wheat phytase, which in turn hydrolyzed 78% of soy-milk phytate in 8 h at 60°C, as compared to 42% hydrolysis observed for the native enzyme under the same conditions (Khare *et al.*, 1994b).

Polyvinyl alcohols (PVA) are water-soluble synthetic polymers with excellent film-forming, adhesive, and emulsifying properties and outstanding resistance to oil, grease and solvents. Chemically, PVA can be considered a polyhydric alcohol with secondary hydroxyl groups on alternate carbon atoms. Although PVA is usually used industrially without chemical modification, it undergoes a reaction typical of long-chain polyhydric alcohols (Davidson, 1980). To modify the structure of phosphorylated PVA gel beads, towards improving their gas permeability, some additives, such as soluble starch, partially saponified PVA, and calcium alginate, were used. A small amount of calcium alginate was added to the gel solution during gelation. The calcium alginate was later dissociated by treating the beads with phosphate solution, thereby effectively promoting gas permeation. The modification process with alginate did not cause a significant decrease in the nitrate reduction rate of the immobilized beads, i.e. the denitrifying activity of the gel beads as determined by the consumption of mg NO_3-N per hour per gram of immobilized sludge. The gas permeability of the beads increased by as much as 62% (Chen *et al.*, 1996).

Another report describes the use of stable, semi-permeable polyamide microcapsules for the encapsulation of baker's yeast. The encapsulated cells were used to reduce 1-phenyl-1,2-propanedione to 2-hydroxy-1-phenyl-1-propanone, a reaction which was performed more efficiently than by free cells or the same preparation encapsulated in κ-carrageenan beads (Green *et al.*, 1996). Glucoamylase was immobilized onto activated poly(2-hydroxyethyl methacrylate)/ethylene glycol dimethacrylate micro-spheres. The size of the micro-sphere had a considerable influence on enzyme activity, i.e. the smaller the sphere, the greater the activity. In 120 h, only 9.0% of the immobilized glucoamylase activity was lost (Arica *et al.*, 1998).

6.2.4 Immobilization by chitosan, low-methoxy pectin and konjak

Chitosan is a polycation that can be used for cross-linking by multi-valent anions. Many reports describing cell entrapment within chitosan beads can be found (Kluge *et al.*, 1982; Stocklein *et al.*, 1983; Vorlop & Klein, 1981). The higher stability of

Fig. 6.3 Hydrocolloid beads of chitosan (magnification ×7).

chitosan beads versus alginate was reported by Stocklein *et al.* (1983). Chitosan beads dissolve in the absence of phosphate and swell under acidic conditions, but are stable in the presence of chelating agents and monovalent cations. Many examples of using chitosan as an immobilization medium have been reported (Fig. 6.3). A few are quoted here. β-Glucosidase was adsorbed on chitosan and cross-linked with glutaraldehyde. The enzyme exhibited a considerable affinity to chitosan, giving good immobilization yields (55–85%) while maintaining an optimum level of activity (550–850 U g^{-1}). The stability in acid buffer ($t_{1/2} > 80-100$ days) and in wine ($t_{1/2} > 55$ days) was also high. The results are promising for future technological application of the immobilized enzyme in wine-making (Martino *et al.*, 1996). Another method for evaluating lactose hydrolysis in a fixed-bed reactor with β-galactosidase immobilized on chitosan has been reported (Carrara & Rubiolo, 1997).

Lactic acid bacteria were micro-encapsulated in cross-linked chitosan membranes formed by emulsification/inter-facial polymerization. Cell viability and activity were demonstrated by the acidification of milk (Groboillot *et al.*, 1993). Another example of immobilization of low-molecular-weight compounds in complex coacervate capsules consisting of water-soluble chitosan salts or acid-soluble chitosan cross-linked with alginate of κ-carrageenan has been reported (Pandya & Knorr, 1991). Enhanced release was detected at 0.5% chitosan. The presence of alginate in the coacervate caused a lower rate of release, whereas the presence of κ-carrageenan in the capsules yielded a faster rate (Pandya & Knorr, 1991).

A DM (degree of methoxylation) of 50% divides commercial pectins into high-methoxyl pectins (HMP) and low-methoxyl pectins (LMP). LMP gels vary in rheological and other properties with calcium content. At low calcium concentrations, LMP gels are soft, coherent and almost transparent. As the calcium concentration increases, the LMP gels become harder, more brittle and less transparent (Davidson, 1980). LMP beads can be produced very easily in a manner similar to alginate beads and can serve for many immobilization purposes

(Gemeiner *et al.*, 1991; Kurillova *et al.*, 1992). LMP and chitosan can be used separately for bead creation; however, due to their potential reaction, their complexes were also used for immobilization (Ullrich *et al.*, 1990). Another report mentions chitosan-coated alginate beads (Zhou *et al.*, 1998).

Konjak flour, which contains a glucomannan called konjak mannan (KM), forms a gel after mild alkali treatment and heating. The gel is liquefied easily at cold temperatures (less than or equal to 10°C). A two-phase dispersion process has been adapted which allows the preparation of spherical konjak beads (50–500 μm diameter). Encapsulation of proteolytic enzymes in cold-melting hydrogel has been investigated (Perols *et al.*, 1996). The objective was to enhance proteolysis during cheese production by maintaining enzymes in the beads during the clotting step and by releasing them during ripening through bead liquefaction (as temperature decreases). The encapsulation yield of Protease B500 was about 50% based on residual proteolytic activity. A low leakage of enzyme at 30°C was found. This makes the system suitable for use during gel formation in cheese manufacture; however, at 4°C, the liberation of enzyme is 7% within 24 h, which is too low to ensure fast cheese ripening. Syneresis of the konjak gel beads under shear stress explains this low enzyme-release rate. Other gels were evaluated (Perols *et al.*, 1996).

6.2.5 Cellulose for entrapment

Cellulose itself is not soluble in water, but it can be dissolved in organic liquids to form beads entrapping bacteria and enzymes (Linko *et al.*, 1977). Cellulose acetate had been used for entrapment, but leakage and diffusional restrictions were reported. A detailed discussion (Dinelli, 1972) of the entrapment of both enzymes and cells in solid fibers made from a variety of polymers, including celluloses, was published in terms of the advantage of using standard equipment for the wet spinning of fibers, a method which can be easily scaled up to any desired quantity.

6.3 Mechanical properties of immobilization matrices

The importance of rheological parameters in assessing the quality of immobilized preparations has been recognized (Lacroix *et al.*, 1990). A few reports have discussed the correlation between chemical and physical properties of beads and their use as immobilization material (Martinsen *et al.*, 1989). The mechanical properties of gel beads are very important in assessing a preparation's stability and operational success. Important parameters are total fracture energy, stress and strain at fracture, and abrasion of gel beads. No correlation between abrasion rate and fracture properties has been observed and abrasion is more likely to be related to fatigue of gel materials (Van Vliet *et al.*, 1997).

Although immobilization in gel matrices is common practice, there seems to be little or no information on how the properties (number, size, etc.) of the entrapped microorganisms influence the mechanical characteristics of the matrix. Some insight has been gained into the influence of bacteria, yeast and fungal spores (Nussinovitch *et al.*, 1994; Nussinovitch, 1994). *Lactococcus lactis* bacteria, *S. cerevisiae* yeasts and *Trichoderma viride* fungal spores were immobilized in agar and alginate gels and the mechanical properties of these gels were tested by compression using a Universal Testing Machine (UTM). Data was collected as volts versus time, and later converted to stress versus strain. Both gel systems exhibited similar behavior after incorporation of the microorganisms. In most cases, the addition of microorganisms up to 10^5 Colony Forming Units ml^{-1} (CFU ml^{-1}) of gel did not change the gel's strength (stress at failure), or affect its deformability modulus or Hencky's strain at failure. A different behavior was observed for lactic acid bacteria in an alginate gel, when 10^7–10^9 cells were immobilized per milliliter of gel matrix (Nussinovitch *et al.*, 1994). Both agar and alginate gels suffered weakened textures (even though their mechanical properties, gelation mechanisms and immobilization methods were very different). Gel strengths and deformability moduli decreased by a factor of 1.2–4.4 (Fig. 6.4) depending on the type of gel, the number of microorganisms per milliter and the diameter of the embedded microorganisms. Following immobilization, gels became more brittle, as demonstrated by a decrease in their Hencky's strain at failure (Nussinovitch *et al.*, 1994).

Insight into the conditions under which the entrapped microorganisms are embedded within the matrix may give some idea of the resultant changed mechanical properties of the entrapping medium with time. When *S. cerevisiae* cells were inoculated at low density in alginate gel beads (Fig. 6.5) and cylinders, cells grew in the form of distinct micro-colonies throughout the gel matrix. Alginate-gel beads gave rise to micro-colonies which became elongated and lens-shaped with their major axes aligned with the gel surface. The aspect ratio (major axis : minor axis length) of the micro-colonies and the local concentration of alginate increased with increasing distance from the center of the gel particles. In contrast, spherical micro-colonies were observed in alginate cylinders formed by internal gelation and no significant local concentration gradients of alginate were detected in these gels. Non-spherical micro-colonies were also observed in carrageenan gel beads. However, the colonies were irregularly shaped, and their major axes demonstrated no preferential alignment (Walsh *et al.*, 1996a). Because the situation of particles embedded within a continuous matrix very much resembles that of a composite material, the question was raised as to whether mathematical models designed for composite materials could also describe gels with entrapped microorganisms. To answer this, *T. viride* fungal spores were immobilized in alginate gels. The mechanical properties of these gels were tested by compression in the UTM. A simple mathematical model for composite materials was found to properly describe the relative stress at failure versus the phase volume of the fungal spores. Random loose packing of the spores was proposed (Nussinovitch, 1994).

Another report studied the stability of alginate and agarose gels used in cell encapsulation (Shoichet *et al.*, 1996). The gel strength of agarose diminished in the presence of cells, because they interfered with the hydrogen-bond formation required for agarose gelation. The gel strength of calcium- or barium-cross-linked alginate, in this case, decreased over 90 days, with an equilibrium gel strength being reached after 30 days. The stability of calcium-cross-linked sodium alginate gels over a 60-day

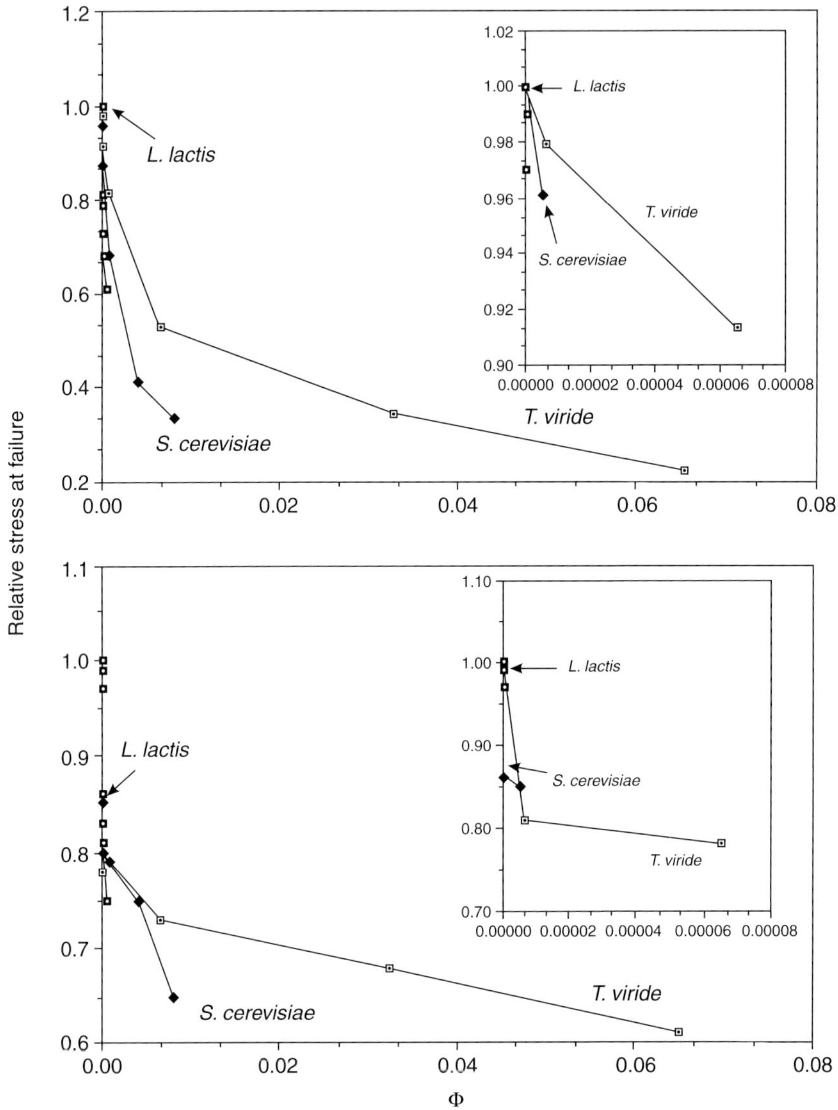

Fig. 6.4 Relative stress (σ) at failure versus the filler (microorganism) volume fraction (ϕ) for bacteria, yeasts and mold spores in agar gels (top) and alginate gels (bottom). Inset, dependency of relative strength at failure on very small occupied volume fractions ($0-8\times10^{-5}$). (From Nussinovitch *et al.*, 1994.)

period was monitored by diffusion of proteins ranging in molecular masses from 14.5 to 155 kD. From these diffusion measurements, the average pore size of the calcium-cross-linked alginate gels was estimated, by semi-empirical model, to increase by 1.76 to 2.89×10^{-8} m over a period of 60 days (Shoichet *et al.*, 1996). The effective diffusion coefficient, D_e, and the distribution constant, K_l, for selected mono- and di-saccharides and organic acids were determined in homogeneous calcium alginate gels with and without entrapped bacteria. The distribution constant (K_l) can be determined from the rate of substrate hydrolysis with and without inhibitor. It is

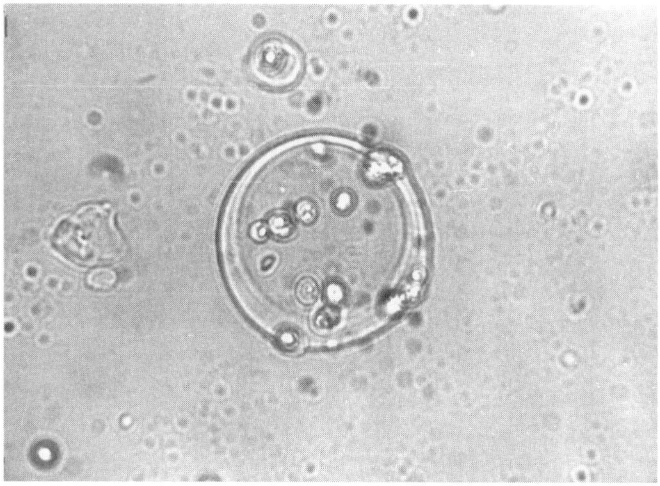

Fig. 6.5 Yeasts immobilized within an alginate bead.

calculated by dividing the final inhibitor concentration by the outcome of subtracting one from the ratio of substrate hydrolysis without inhibitor to hydrolysis with inhibitor. Results were obtained from transient concentration changes in well-stirred solutions of limited volume, in which the gel beads were suspended. The effective diffusion coefficients and the distribution constants were estimated by fitting mathematical model predictions to the experimental data using a non-linear model-fitting program. Both single solute diffusion and multiple solute diffusion were performed (Oyaas *et al.*, 1995). A small positive effect was obtained on the values of D_e for the system of multiple solute diffusion; however, the values of K_t were not significantly influenced. For the nine solutes tested, D_e for 2% calcium alginate gel beads was found to be approximately 85% of the diffusivity measured in water. The effects on D_e and K_t of lactose and lactic acid were determined for a variety of alginate concentrations, pHs, temperatures, and biomass contents in the beads. D_e decreased linearly for both lactose and lactic acid with increasing cell concentration in the calcium alginate gel (Oyaas *et al.*, 1995). K_t was constant for both lactose and lactic acid with increasing cell concentration. D_e was significantly lower at pH 4.5 than at pH 5.5 or 6.5 for both lactose and lactic acid. Furthermore, D_e seemed to decrease with increased alginate concentration in the range of 1–4%. The diffusion rate increased with increasing temperature, and the activation energy for the diffusion process for both lactose and lactic acid was constant in the temperature range tested (Oyaas *et al.*, 1995).

6.4 Mixed systems for immobilization

Mixed systems sometimes produce better solutions for immobilization. A mixed alginate–gelatin system for yeast encapsulation was studied. Following bead formation, the alginate was leached out in phosphate buffer and the porous matrix produced by the gelatin was stabilized by cross-linking with glutaraldehyde (SivaRaman *et al.*, 1982). As previously described, alginate alone can form a capsule, but more sophisticated applications have been proposed by combining alginate with the negatively charged polyamino acid polylysine. In the latter case, the barrier is a weak polyelectrolyte complex formed by the reaction of alginate and polylysine. In many cases, an extra layer of alginate is applied to the preformed membrane to achieve better compatibility. Experiments were also carried out to assess the utility of coating alginate spheres entrapping biomass with a polyacrylamide resin (Eudragit RL100). The studied properties included mechanical resistance, biomass leakage and mass-transfer resistance. In addition, bioactivity tests were carried out using entrapped *S. cerevisiae*. The mechanical resistance of the coated beads was higher and the biomass leakage lower than with non-coated controls, while the bioactivity and the mass-transfer resistance were similar. Coated beads may therefore be of use in bioreactor systems (Ruggeri *et al.*, 1991). Other mixed systems include complex coacervate capsules or fibers with double gel layers. Coacervation involves the separation of a liquid phase of coating material from a polymeric solution, followed by the coating of that phase as a uniform layer around suspended core particles. The coating is then solidified. In general, the batch-type coacervation processes consist of three steps: formation of a three-dimensional immiscible chemical phase, deposition of the coating and solidification of the coating, all carried out under continuous agitation (Pegg & Shahidi, 1999). Such immobilization preparations sometimes provide extra stability for batch reuse, or the ability to produce special products. An example could be the immobilization of plant cell culture in complex coacervate capsules for secondary metabolite production (Knorr *et al.*, 1990), or a co-immobilized culture system of (Citr+) *L. lactis* subsp. *lactis* 3022 cells (outer layer) and homogenized bovine liver (inner layer) as the growth substrate for the bacteria, in calcium alginate fibers with double gel layers. The culture system gave high diacetyl productivity (greater than or equal to 30 mg l^{-1}) for ten repeat-batch cultures (Ochi *et al.*, 1991).

6.5 Food and biotechnological uses of immobilization

6.5.1 Biotransformation from geraniol to nerol

The abilities of two grapevine cell suspensions (Gamay and Monastrell) to biotransform geraniol into nerol in a biphasic system based on culture medium and Miglyol 812 were compared. Geraniol is a fragrant, pale yellow liquid alcohol, $C_9H_{17}COH$, derived from the oils of geranium and citronella and used in cosmetics and flavorings. Nerol is a colorless liquid, derived from orange blossoms and used in perfumery. Geraniol and nerol are geometric (*cis–trans*) isomers. Geraniol has a waxy, sweet rose odor; nerol also smells of sweet rose with a citrus tang. The Gamay grape-cell suspension was able to transform a higher concentration of geraniol into nerol than the Monastrell one. Immobilization proved to be advantageous in protecting cells against the toxicity of the substrate. Furthermore, immobilization seemed to have an effect on secondary metabolism: the cells immobilized in polyurethane foam were more efficient at performing the immerization process than either the freely suspended or calcium-alginate-immobilized cells (Guardiola *et al.*, 1996).

6.5.2 *Bioartificial organs*

Current membrane-based bioartificial organs consist of three basic components: a synthetic membrane, cells that secrete the product of interest and encapsulated matrix material. Alginate and agarose have been widely used to encapsulate cells for artificial organ applications. It is important to understand the degree of transport resistance imparted by these matrices in cell encapsulation to determine whether adequate nutrient and product fluxes can be obtained (Li *et al.*, 1996). In general, 2–4% agarose gels offered little transport resistance for solutes of up to 150 kD, whereas 1.5–3.0% alginate gels offered significant transport resistance for solutes in the range of 44–155 kD, lowering their diffusion rates by 10–100-fold as compared to their diffusion in water. Doubling the alginate concentration had a more significant effect on hindering the diffusion of larger-molecular-weight species than did doubling the agarose concentration. Average pore diameters were approximately 1.70×10^{-8} and 1.47×10^{-8} m for 0.5 and 3% alginate gels, respectively, and 4.80×10^{-8} and 3.60×10^{-8} m for 2 and 4% agarose gels, respectively. These values were estimated using a semi-empirical correlation based on diffusional transport of different-size solutes. This method, developed for measuring diffusion in these gels, is highly reproducible and useful for gels cross-linked in the cylindrical geometry relevant for studying transport through matrices used in cell immobilization in the hollow-fiber configuration (Li *et al.*, 1996).

6.5.3 *Organic acid fermentation and conversion*

Fruit ripening is accompanied by changes in organic acids. These reach the maximum during the growth and development of the fruit on the tree, but decrease during storage; these changes are also highly dependent on temperature. The Krebs cycle is active in cells of higher plants in generating a variety of organic acids, including citric, malic and succinic acids. Citric and malic acids are important constituents of most fruits, with oranges, lemons, and strawberries being high in citric acid, and apples, pears, and plums high in malic acid (Eskin, 1990). The use of organic acids in basic foods and food mixtures is on the rise. Different uses include: in cream soups, custard mixtures and pies, gels, pastes, dressings, cakes and pickling solutions, among many others. Many examples of immobilization for the production of organic acids from different raw materials can be found in the literature (Gu, 1997).

Citric acid ($C_6H_8O_7$, MW 192.13, pK 3.14) is used in many foods. It can be utilized as a sequestrant, and is generally recognized as safe when used in accordance with good manufacturing practice (Doores, 1990). Citric acid is one of the major acidulants in carbonated beverages, imparting a tangy citrus flavor. It is commercially used as a synergist for antioxidants and as a retardant of browning reactions. Other uses are as a plasticizer and emulsifier to provide texture to processed cheese and to enhance melting; to reduce heat-processing requirements by lowering the pH; and to control acidity in gel formation (Doores, 1990).

Citric acid can be produced from glucose by calcium alginate-immobilized *Yarrowia lipolytica* yeast. The highest acid concentration ($39\,g\,l^{-1}$) was produced in a medium consisting of $150\,g\,l^{-1}$ glucose, $0.105\,g\,l^{-1}$ potassium hydrogen phosphate, $0.84\,g\,l^{-1}$ magnesium sulfate and $21\,mg\,l^{-1}$ copper sulfate. Hardening the beads and activating the immobilized biocatalyst in a nutrient solution can improve these results (Rymowicz *et al.*, 1993). Another study reported the production of citric acid by immobilized *Y. lipolytica* yeast in repeat-batch, shaken-flask and airlift fermentation. The highest citric acid production was observed in alginate beads versus other carriers based on κ-carrageenan, polyurethane gel and foam, and nylon web; the smaller the bead size, the greater the productivity (Kautola *et al.*, 1991). Citric acid, oxalic acid, erythritol and glycerol formation by *Aspergillus niger* immobilized in calcium alginate was reported in another report. Morphological changes were strain-dependent (Hamdy *et al.*, 1992). Agarose beads containing *A. niger* were also used to produce citric acid from soy whey. Maximal citric acid yields of 21 and $27\,g\,l^{-1}$ with free and immobilized cells were reported, respectively (Khare *et al.*, 1994a).

Zymomonas mobilis can convert glucose and fructose to gluconic acid and sorbitol. The enzyme, glucose–fructose oxidoreductase, catalyzing the intermolecular oxidation–reduction of glucose and fructose to gluconolactone and sorbitol, was formed in high amounts ($1.4\,U\,mg^{-1}$) when *Z. mobilis* was grown in chemostats with glucose as the only carbon source under non-carbon-limiting conditions. The activity of a glucono-lactone-hydrolyzing lactonase was constant at $0.2\,U\,mg^{-1}$. Using glucose-grown cells for the conversion of equimolar fructose and glucose mixtures of up to 60% (w/v), a maximum product concentration of only $240\,g\,l^{-1}$ of sorbitol was found (Rehr *et al.*, 1991). The accumulated gluconic acid was further metabolized to ethanol. After permeabilizing the cells using cationic detergents, maximum sorbitol and gluconic acid concentrations of $295\,g\,l^{-1}$ each were reached; no ethanol production occurred. In a continuous process with κ-carrageenan-immobilized and polyethylenimine-hardened, permeabilized cells, no significant decrease in the conversion yield was observed after 75 days. The specific production rates for a high-yield conversion (>98%) in a continuous two-stage process were $0.19\,g\,l^{-1}\,h^{-1}$ for sorbitol and $0.21\,g\,l^{-1}\,h^{-1}$ for gluconic acid. For the sugar conversion of cetyltrimethyl-ammonium bromide-treated κ-carrageenan-immobilized cells, a V_{max} of $1.7\,g\,l^{-1}\,h^{-1}$ for sorbitol production and a Km of $77.2\,g\,l^{-1}$ were determined (Rehr *et al.*, 1991). Another report (Kim & Kim, 1992) described the simultaneous production of gluconic acid and sorbitol from glucose and Jerusalem artichoke using a glucose–fructose oxidoreductase of *Z. mobilis* and inulinase. Inulinase was immobilized on chitin by cross-linking with glutaraldehyde. Cells of *Z. mobilis* permeabilized with toluene were co-immobilized with chitin-immobilized inulinase in alginate beads. The optimum amounts of both chitin-immobilized inulinase and permeabilized cells for co-immobilization were determined, and operational conditions were optimized. In a continuously stirred tank reactor operation, the maximum productivities for gluconic acid and sorbitol were about 19.2 and $21.3\,g\,l^{-1}\,h^{-1}$, respectively, at a dilution rate of $0.23\,l\,h^{-1}$ and a substrate concentration of 20%, but operational stability was low because of bead abrasion (Kim & Kim, 1992). As an approach to increasing the operational stability, a recycling packed-bed reactor (RPBR) was employed. In RPBR operation, the maximum productivities for gluconic acid and sorbitol were 23.4 and 26.0 $g\,l^{-1}\,h^{-1}$, respectively, at the dilution rate of 0.35 $l\,h^{-1}$ and a substrate concentration of 20% when the re-circulation rate was fixed at 900 $ml\,h^{-1}$. Co-immobilized enzymes were stable for 250 h in an RPBR without any loss of activity, while half-life in a continuously stirred tank reactor was about 150 h (Kim & Kim, 1992).

Lactic acid ($C_3H_6O_3$, MW 90.08, pK 3.08) is used extensively for its sensory qualities. It is also used as an antimicrobial agent, pH-control agent, curing and pickling agent, flavor enhancer, flavoring agent and adjuvant, solvent and vehicle (Doores, 1990). Lactic acid was continuously produced from raw starch using a combination of a reversibly soluble-autoprecipitating amylase (D-AS) depending on pH and *Lactobacillus casei* entrapped in κ-carrageenan. Lactic acid was also produced continuously in a novel reactor system, consisting of a turbine-blade reactor with a cylindrical stainless-steel net, a mixing vessel, and a separation vessel. The gel beads with entrapped lactic acid bacteria were held on the cylindrical net in the main reactor throughout cultivation. D-AS was separated continuously from a solution containing lactic acid by self-sedimentation in the separation vessel, and it was returned to the main reactor for repeated use (Hoshino *et al.*, 1991). In the continuous lactic acid production from raw starch, the lactic acid productivity was 3.1 $g\,l^{-1}\,h^{-1}$ at a dilution rate of 0.11 h^{-1} and the value was about 3.1 times higher than the average of the repeat batches of lactic acid production. Although D-AS becomes inactivated due to insolubilization of the enzyme by KCl accumulated during the control of pH in the reactor, it is possible to recover the enzymatic activity by replacing some of the old broth with new one. This continuous production system, using the novel reactor, may be widely applicable to the production of useful materials from solid substrates with microorganisms other than lactic acid bacteria for fermentation (Hoshino *et al.*, 1991).

Production of L(+)-lactic acid by *Rhizopus oryzae* immobilized in alginate in a tapered-column fluidized-bed batch reactor has also been reported (Hamamci & Ryu, 1994). Immobilized *Sporolactobacillus inulinus* ATCC 15538 cells were used to produce D(−)-lactic acid from topinambour tubers in a stirred reactor. The biocatalyst retained its initial activity level for 400 h of operation at 35°C providing lactic acid yields of 95–96% (Abelyan & Abelyan, 1996).

Alginate was also favored for the immobilization of *L. casei* for lactate production under stirred-batch, as well as packed-bed conditions (Dong *et al.*, 1991). However pH control was a problem in the packed-bed reactor. Stirred-batch experiments, terminated with total glucose utilization, yielded 90–99% lactate and a total productivity of 1.6 $g\,l^{-1}\,h^{-1}$ (Dong *et al.*, 1991). Different beads produced from κ-carrageenan (2.75% w/w)–LBG (0.25% w/w) have been used to entrap *L. casei* and effective diffusion coefficients and equilibrium partition factors for lactose and lactic acid were determined. The use of mathematical models of non-steady state diffusion into and/or from a sphere and appropriate boundary conditions as a tool for the calculation of the diffusion coefficients were described (Arnaud *et al.*, 1992).

A continuous bi-particle fluidized-bed reactor (BFBR) has been developed for the simultaneous fermentation and recovery of lactic acid. In this processing scheme, bacteria are immobilized in gelatin beads and are fluidized in a columnar reactor. Solid particles (weak-base resin IRA-35) with sorbent capacity for the product are introduced at the top of the reactor and fall counter-currently to the biocatalyst, effecting in situ removal of the inhibitory lactic acid while also maintaining reactor pH at optimal levels. Fermentation trials of 1 week using immobilized *Lactobacillus delbreuckii* with sorbent addition demonstrated a volumetric productivity (6.9 $g\,l^{-1}\,h^{-1}$) at least 16-fold higher than that of a free-cell batch fermentation with base pH control and identical biomass concentration and medium composition (Kaufman *et al.*, 1996). Regeneration of the loaded sorbent from the BFBR yielded a concentration of lactic acid that was 35-fold original levels in the fermentation broth (70 versus 2 $g\,l^{-1}$). Lactic acid concentrations as high as 610 $g\,l^{-1}$ were observed when the loading solution contained 50 $g\,l^{-1}$ lactic acid. Rich medium formulations did not seem to increase BFBR performance. The benefits of this reactor system, as opposed to conventional batch fermentation, are discussed in terms of productivity and process economics (Kaufman *et al.*, 1996). Calcium alginate beads were used as a carrier for the immobilization of *L. casei* and *L. lactis* for continuous production of lactic acid from de-proteinized whey. Average lactic acid productivity, yield, and lactose utilization were 24 $g\,l^{-1}\,h^{-1}$, 55% and 90%, respectively (Roukas & Kotzekido, 1996).

Aspergillus oryzae grown in situ from spores entrapped in calcium alginate gel beads was used for the production of kojic acid. Kojic acid – 5-hydroxy-2-(hydroxymethyl)-4H-pyran-one ($C_6H_4O_4$) – is a form of bleaching agent which is rather effective at reducing pigmentation irregularities. The immobilized cells in flask cultures produced kojic acid in a linear proportion while maintaining the stable metabolic activity for prolonged production. Kojic acid was accumulated up to 83 $g\,l^{-1}$, at which point the kojic acid began to crystallize, and, thus, the culture had to be replaced with fresh media for the next batch culture. The overall productivities of two consecutive cultivations were higher than that of free mycelial fermentation. However, the production rate of kojic acid by the immobilized cells was suddenly decreased with the appearance of central cavernae inside the immobilized gel beads after 12 days of third-batch cultivation (Kwak & Rhee, 1992).

The yeast *S. cerevisiae* was amplified for the enzyme fumarase by cloning the single nuclear gene downstream of a strong promoter. The over-producing strain converted fumaric acid to L-malic acid at a rate of 65 $mM\,g^{-1}\,h^{-1}$ in free-cell experiments, and approximately 87% of the fumaric acid was converted to L-malic acid within 45 min (fumaric acid is $C_4H_4O_4$; malic acid is $C_4H_6O_5$). Fumaric acid has been used as an antimicrobial agent, in the prevention of malolactic fermentation in wines as well as a means of adding acidity to wines. Malic acid can be used as a flavor enhancer, flavoring agent and adjuvant and pH control agent. Activity was dependent on the addition of surfactant to the medium, and minimal activity was seen with the wild-type yeast strain. The constructed strain was immobilized in agarose beads (2.4 mm mean diameter) and within agarose micro-spheres (193 and 871 μm mean diameter). The rate of bioconversion increased with decreasing bead diameter, with similar rates from the 193 μm diameter micro-spheres and the free cells. The presence of surfactant was essential for initial activity of the immobilized cells; however, high activity was observed in subsequent experiments in the absence of surfactant. Stable activities over a 48 h period were maintained within the large-diameter agarose beads, while decreasing activities were observed within the agarose micro-spheres (Neufeld *et al.*, 1991). L-Malate was produced from fumarate by using immobilized *S. cerevisiae* cells entrapped in polyacrylamide. This preparation performed better when pre-treated with malonate. Under the experimental conditions described here, succinate was not detected as a by-product of the reaction, as had been reported for other micro-organisms (Figueiredo & Carvalho, 1991). Fumarate bioconversion to L-malic acid by *S. cerevisiae* entrapped within polyacrylamide gel beads has been described (Oliveira *et al.*, 1994). The spherical configuration of the beads was improved by the addition of detergent (sodium dodecyl sulfate) addition. Bioconversion was 60 times higher than that found with free cells (Oliveira *et al.*, 1994). Other reports deal with the effects of propionic acid on propionibacteria fermentation (Zhong *et al.*, 1998), and improved organic acid production by calcium alginate-immobilized propionibacteria (Rickert *et al.*, 1998).

Ingredients can be produced from acids by immobilization preparations (Shiow *et al.*, 1998). *Sporidiobolus salmonicolor* CCRC 21975 was immobilized in κ-carrageenan, chitosan, agarose or calcium alginate. Due to the detrimental effects of the high temperature attained during the gelling of κ-carrageenan and agarose, as well as the toxicity of chitosan to the test organism, immobilization of *S. salmonicolor* with these matrices for the production of γ-decalactone was inadequate. γ-Decalactone, $C_{10}H_{18}O_2$, is a colorless to pale yellow liquid with a fruity, peach-like odor. Neither viable cells nor production of γ-decalactone could be detected in media after 4 days cultivation of *S. salmonicolor* immobilized with κ-carrageenan or chitosan. Fewer viable cells and little γ-decalactone production were found in media with agarose-immobilized cells. In contrast, no significant reduction in the viable population was noted during immobilization procedures using alginate. Alginate-immobilized *S. salmonicolor* cells showed less susceptibility to ricinoleic acid toxicity and produced more γ-decalactone than did free cells. Time courses of γ-decalactone production by *S. salmonicolor* also revealed that immobilized cells produced a maximum γ-decalactone yield of ca. $131.8\,mg\,l^{-1}$ after 5 days fermentation, compared with a maximum of ca. $107.5\,mg\,l^{-1}$ for free cells (Shiow *et al.*, 1998).

Another food-related application is the bioconversion of vanillin (widely occurs in nature, a synthetic flavoring with the empirical formula: $C_8H_8O_3$, a white to yellowish crystalline powder with a characteristic strong, vanilla-like odor; very sweet taste) into vanillic acid. Alginate beads served as carriers for *Pseudomonas*, transformation rate was only 47% after 13 h of conversion. However, optimal yield of higher than 80% has been detected in a continuous immobilized cell reactor during 76 h of operation (Bare *et al.*, 1992).

6.5.4 Ethanol, wine and vinegar

Alcohols can be used in foods as carriers for additives, confection glazes, ice creams, ices, liquors, sauces and sprayable vegetable oils. Ethanol can be produced by fermentation of xylose and rice straw hydrolysate using free and immobilized cells of *Candida shehatae* (NCL-3501) under batch, fed-batch and continuous-culture conditions. In general, ethanol yields (g ethanol per g sugar utilized) were higher with immobilized cells in all reactor types. Fed-batch or continuous cultures exhibited higher ethanol yields and volumetric productivities. *C. shehatae* efficiently utilized the sugars present in rice straw hemicellulose hydrolysate, prepared by two different methods, within 48 h. In continuous culture, using acid hydrolysate of rice straw, a steady state could be maintained for 216 h with a volumetric ethanol productivity of 0.33 $g\,l^{-1}$ (Abbi *et al.*, 1996).

In modern wine-making practice, crushing and stemming are carried out mechanically. The resultant juice (with or without skin and pips), referred to as *must*, is treated with sulfur dioxide to inhibit wild yeast and spoilage bacteria (Sutherland *et al.*, 1986). The possibility of using immobilized laccase (EC 1.10.3.1.) for the enzymatic removal of phenolic compounds from must and wine has been tested on a laboratory scale. The immobilized enzyme was packed in a column and model solutions of catechin or white grape must were eluted through. Oxygen saturation, flow rate and immobilized enzyme amount were the main parameters used to study phenolic oxidation (Brenna & Bianchi, 1994). Semi-continuous and continuous riboflavin production by calcium alginate-immobilized *Candida tropicalis* in concentrated, rectified grape must has also been reported (Buzzini & Rossi, 1998).

A method based on the survival of yeast cells subjected to ethanol or heat shock was utilized to compare the stress resistance of free and carrageenan-immobilized yeast cells. Results demonstrated a significant increase in yeast survival against ethanol for immobilized cells as compared to free cells, while no marked difference in heat resistance was observed. When entrapped cells were released by mechanical disruption of the gel beads and submitted to the same ethanol stress, they exhibited a lower survival rate than entrapped cells, but a similar or slightly higher survival rate than free cells. The incidence of ethanol- or heat-induced respiratory-deficient mutants of entrapped cells was equivalent to that of control or non-stressed cells ($1.3 \pm 0.5\%$), whereas ethanol- and heat-shocked free and released cells exhibited between 4.4% and 10.9% average incidence of respiration-deficient mutants. It was concluded that the carrageenan gel matrix provides protection against ethanol, and that entrapped cells return to normal physiological behavior as soon as they are released. Cell growth rate was a significant factor in the resistance of yeast to high ethanol concentrations. The optimum conditions for reliable and reproducible results involved the use of slow-growing cells after exhaustion of the sugar substrate (Norton *et al.*, 1995). Kissiris, γ-alumina and calcium alginate were the supports for immobilization of a cryo-tolerant and alcohol-resistant strain of *S. cerevisiae*. Continuous wine-making with immobilized cells on each of these supports was performed at ambient and low temperatures. To evaluate the results of continuous wine-making with immobilized cells, batch fermentations were performed using immobilized and free cells separately, in the same range of temperatures. Fourfold higher ethanol productivities at room temperature and 10-fold higher productivities at low temperature were obtained by continuous wine-making, in comparison to batch fermentations performed with free cells (Bakoyianis *et al.*, 1997). Specifically, at 7°C, ethanol productivities achieved by continuous wine-making were

16.7, 13.8 and 23.2 $g\,l^{-1}$ day^{-1}, and by batch wine-making were 4.5, 5.1, and 5.6 $g\,l^{-1}$ day^{-1} respectively, for kissiris, γ-alumina and calcium alginate. For free cells, ethanol productivity was 1.5 $g\,l^{-1}$ day^{-1} at 7°C. The three continuous wine-making systems were operated continuously for 80 days without any infection or decrease in ethanol productivity. Moreover, the wines were produced with low total and volatile acidities (Bakoyianis *et al.*, 1997).

Traditional sparkling wines are produced by secondary fermentation. After the main fermentation is complete, sugar and yeasts are added to produce a secondary fermentation with an excess of carbon dioxide. This may take place in the bottle, as with champagne, or in closed vats. Sparkling wines may also be produced without secondary fermentation by carbon dioxide injection (Sutherland *et al.*, 1986). Production of bottle-fermented sparkling wine using yeast immobilized in double-layer gel beads or strands has been reported (Yokotsuka *et al.*, 1997). Bottle-fermented sparkling wines were produced using *S. cerevisiae* immobilized within a double layer of calcium alginate beads or strands, and factors affecting the leakage of viable cells from the gel into the wine during fermentation and ageing were investigated. Yeast immobilized in beads or strands at 10^4, 10^6 or 10^8 cells ml^{-1} was added to 740-ml samples of a base wine containing 24 $g\,l^{-1}$ sucrose, at ethanol concentrations of 0.5, 3, 6, 9 or 12% (v/v). Secondary fermentation was conducted at 15 or 25°C in 770-ml bottles with a pressure gauge (Yokotsuka *et al.*, 1997). Fewer free yeast cells appeared in the wines with higher initial numbers of immobilized cells or the initial ethanol concentration. Wines fermented with yeast immobilized in gel beads contained greater numbers of yeast cells than wines with yeast immobilized in gel strands, but no free viable yeast cells remained in the wines a few months after fermentation in either case. Beads are preferable to strands in commercial production because they are much more easily added to and removed from bottles via the ice-disgorging procedure commonly used in the production of champagne, without the need for traditional riddling. Sparkling wine has also been made using freely suspended yeast, and changes in chemical components, including amino acids, during ageing of wines made with free and immobilized yeast were investigated and compared. There were no significant chemical differences between the two. It was thus concluded that secondary fermentation of sparkling wines using yeast immobilized within double-layer alginate beads is practical for commercial production (Yokotsuka *et al.*, 1997).

Vinegars may be produced from any raw material containing sufficient sugar or alcohol. Examples are fruit juices, starchy vegetables (potato, sweet potato), malted cereals, sugars (molasses, honey), and alcoholic beverages or dilute ethanol. Vinegar manufacture is a two-stage process, the production of alcohol from carbohydrate and the subsequent oxidation of alcohol to ethanoic acid (Sutherland *et al.*, 1986). The optimum conditions for *Acetobacter* immobilization were investigated. The results show that the maximum oxygen uptake rate (OUR_m) and cell release are related to alginate and cell concentration in the gel. Different alginate concentrations did not affect cell viability, but long storage in calcium chloride reduced the number of living cells. The double alginate gel layers had no influence on cell viability or on the OUR_m and prevented cell leakage from the gel matrix (Fumi *et al.*, 1992). *Acetobacter aceti* cells were immobilized using entrapment in a calcium alginate gel and adsorption on preformed cellulose beads. The cell number within the supports showed no significant alterations with changing temperature or pH, whereas acetic acid production was slightly increased by immobilization (Krisch & Szajani, 1996).

6.5.5 Xylitol production

Xylitol is a pentitol that can be found in most fruits and berries as well as vegetables. Commercially, it is produced from xylan-containing plant material by acid hydrolysis, hydrogenation and purification. Xylitol can also be produced by microbiological methods. Xylitol is a sweetener, supplied as a colorless, non-hygroscopic crystal and has a caloric content equal to sucrose. It is non-carcinogenic and can be used for diabetic and dietetic foods (Salminen & Hallikainen, 1990). Many reports have been written on xylitol production (Dominguez, 1998). *Eucalyptus globulus* wood hydrolysates were concentrated by vacuum evaporation to increase their xylose contents, treated with activated charcoal, supplemented with nutrients and used as culture media for xylitol production by *Debaryomyces hansenii* NRRL Y-7426. The susceptibility of hydrolysates to fermentation was strongly dependent on the initial cell concentration. Media containing 58–78 g xylose l^{-1} were hardly consumed in batch experiments starting with 16 g cells l^{-1}, whereas 39–41 g xylitol l^{-1} were yielded by fermentations carried out with a similar concentration of the carbon source and initial cell concentrations of 50–80 g l^{-1} (Parajo *et al.*, 1996).

Aerobic cultures of *S. cerevisiae* produced xylitol, but not ethanol with D-xylose as a sole carbon substrate. In the presence of a glucose and xylose mixture, the consumed xylose was nearly stoichiometrically converted to xylitol. Under anaeorobic conditions the conversion rate decreased. Agar-entrapped yeasts behaved like anaerobically grown cultures, but with higher specific rates of xylitol production (Lebeau *et al.*, 1997). Agar in a special composite immobilized-cell structure, consisting of a flat agar layer between two micro-porous membrane filters, was also reported to serve as a matrix for such a conversion (Lebeau *et al.*, 1998).

6.5.6 Carotenoids and leucrose

Production of a carotenoid-rich product by alginate entrapment and fluid-bed drying of *Dunaliella salina* has been reported. The culture medium was composed of sea water, sea salt, KNO_3 and K_2HPO_4, and CO_2 was bubbled in to maintain culture pH at 8.0 (Leach *et al.*, 1998). Production of leucrose by dextran–sucrase has also been mentioned (Buchholz *et al.*, 1998). Leucrose is a novel, non-cariogenic disaccharide and as such is a highly promising sugar substitute for the prevention of caries. Leucrose possesses excellent nutritional properties with regard to metabolic utilization and is well tolerated (Buchholz *et al.*, 1998).

6.5.7 Immobilization in the milk industry

Whey is the watery, fluid part of the milk that remains after the curd, or casein precipitate, has been removed. Whey contains whey proteins which are not precipitated by either acid or renin, along with most of the lactose and water-soluble vitamins and minerals. Whey is used primarily as animal feed but some is also used in making special cheeses. The use of whey in the production of alcohol and beer has been investigated. It can also be used as a liquid for gelatin desserts and salads (Gates, 1981). The production of lactic acid from de-proteinized whey by mixed cultures of free and co-immobilized *L. casei* and *L. lactis* cells in batch and fed-batch culture was investigated. Fed-batch culture proved to be a better fermentation system for the production of lactic acid than batch culture. The maximum lactic acid concentration ($46\,g\,l^{-1}$) in fed-batch culture was obtained with both the free-cell mixture and co-immobilized cells at a substrate concentration of $100\,g\,l^{-1}$ and a feeding rate of $250\,ml\,h^{-1}$. In repeat fed-batch culture, co-immobilized *L. casei* and *L. lactis* cells gave a higher overall lactic acid concentration compared with the free-cell mixture. The co-immobilized *L. casei* and *L. lactis* cells in calcium alginate beads retained their ability to produce lactic acid for 20 days (Roukas & Kotzekidou, 1998).

Lactobacillus helveticus entrapped in κ-carrageenan–locust bean gum (LBG) gel beads was found to be suitable for continuous lactic acid fermentation of whey permeate enriched with yeast extract. At pH 4.7–6.3, entrapped cells were responsible for 75–85% of the lactic acid and biomass production, and at pH 4.3, production increased to 90% (Norton *et al.*, 1994). An isolate from Egyptian Cheddar cheese has been observed to be an effective lactic acid producer from salt whey permeate to which yeast extract and minerals have been added. Agarose beads were used to immobilize the culture, producing a steady lactic acid concentration of $33.4\,mg\,ml^{-1}$ (Zayed & Winter, 1995). Another report discussed the use of salt whey as a substrate for lactic acid production by *L. casei* entrapped within agar gel beads and this system was operational even at 8% salt. The bead kept its stability and structure for 168 h in the salt medium (Zayed & Zahran, 1991). Other beads produced from κ-carrageenan–LBG gel were used to immobilize three strains of *Lactococcus*, for the continuous fermentation of a whey ultra filtration permeate medium. During the process no strain was eliminated or became dominant and beads kept their integrity and stability (Lamboley *et al.*, 1997). Calcium alginate beads have also been tried for the production of *L. lactis* subsp. *cremoris* in media and growth conditions that have been modified simultaneously. Tested media were whey, whey supplemented with yeast extract and/or meat extract, milk or the commercial medium Gold Complete (Nordica). Biomass production has been proven as an effective method for achieving a concentrated cell suspension without filtration or centrifugation (Morin *et al.*, 1992).

Fermented milk products are defined as those in which the fundamental nature of the product is derived from souring by a culture of lactic acid bacteria. For practical purposes, this means hard and soft cheeses, yogurt and related products (Sutherland *et al.*, 1986). Microbial dynamics of jointly and separately entrapped mixed cultures of mesophilic lactic acid bacteria during the continuous pre-fermentation of milk has been described (Sodini *et al.*, 1997).

During milk fermentation by immobilized *L. lactis* in calcium alginate beads, some cells are released. Re-utilization of the beads resulted in 30 times the number of free cells in the medium. Rinsing the beads between fermentations had no influence on the number of free cells in the milk. Even coating the beads with poly-D-lysine did not significantly reduce the release of cells during five successive fermentations. Double coating was successful to some extent, but acidification of milk with these beads was slower in comparison to non-coated beads. Ethanol treatment and heating of beads killed cells on their peripheral areas and were found to be effective in minimizing release due to cells entrapped near the bead surface (Champagne *et al.*, 1992). Regular and two-layer calcium alginate beads served for entrapment of *L. lactis* subsp. *lactis* and *L. lactis* subsp. *lactis* biovar. *diacetylactis* for fermentation of pasteurized cream (31% fat content). The beads permitted a higher concentration of inoculum followed by a reduction in the fermentation time (Prevost & Divies, 1992).

6.5.8 Amino acid production

Amino acids are the building blocks of proteins. When food is digested, proteins are decomposed to amino acids, which are then absorbed into the blood and carried in the blood stream to participate in the synthesis or formation of new body proteins. Amino acids that cannot be prepared in the body are called essential amino acids. Except for some infant formulas and parental solutions, the use of amino acids as food additives is limited for the most part to the essential ones. Alginate and agar gels have been used to entrap cells of *Brevibacterium* sp. (DSM 20411) for the production of L-glutamic acid. The production medium contained glucose, urea, KH_2PO_4, biotin, a mineral solution, corn steep liquor and one drop of Tween 80 in distilled water. Maximum L-glutamic acid production in batch and repeat-batch modes was 7.4 and 8.7 $mg\,ml^{-1}$, respectively, with alginate beads, and 11.8 and 13.3 $mg\,ml^{-1}$, respectively, with agar beads. Optimized conditions yielded the production of 10–11 $mg\,ml^{-1}$ for 2 weeks in continuous mode, in a packaged column bioreactor with immobilized cells (Nampoothiri & Pandey, 1998). Another manuscript describes prolonged production of tryptophan using immobilized bacteria (Dallmann *et al.*, 1997).

6.5.9 Production of oligosaccharides

Disaccharides (sucrose, maltose, lactose, etc.) are derived by the elimination of one molecule of water from two molecules of monosaccharides. Similarly, polysaccharides are formed by the elimination of $n-1$ molecules of water from n molecules of monosaccharides. The boundary between oligosaccharides and polysaccharides is normally taken as somewhere around degree of polymerization (DP) 20. Fructose oligosaccharides (FOS) can be produced by immobilization of a soil-isolated strain of *Aspergillus japonicus* in calcium alginate beads, for fermentation of a medium composed mainly of sugar cane molasses. Optimum pH and temperature were 5.0–5.6 and 55°C, respectively. The microorganisms and the process utilized appear promising for industrial applications (Cruz *et al.*, 1998). Other reports have concentrated on: production of isomalto/branched oligosaccharide syrup by using immobilized neopullulanase and preliminary evaluation of the syrup as a food additive (Kuriki *et al.*, 1997); production of high-purity FOS with co-immobilized *A. niger* and enzyme (Jiang *et al.*, 1996); production of FOS by immobilized mycelium of *A. japonicus* (Chih *et al.*, 1996). In producing FOS, fructose transformation of sucrose was inhibited by its by-product glucose, thus making it impossible to obtain high-purity FOS: purity was generally 49%. Removal of isomerization of glucose was tested by two processes: converting glucose with glucose oxidase, or use of immobilized glucose isomerase. Glucose oxidase was cross-linked with 2% glutaraldehyde and 0.1% tannin and co-immobilized with *A. niger* in a calcium alginate gel. Co-immobilized beads were incubated at pH 5 and 50°C for 24 h and FOS were produced with 71% purity. In the other process, immobilized *A. niger* and immobilized glucose isomerase were co-packed in a column that produced FOS of 63% purity and 16% fructose. The pH optima were 5 to 5.5 for cell multiplication and 7 for reaction of immobilized isomerase (Jiang *et al.*, 1996).

6.5.10 Preservatives and bacteriocins

Preservatives serve a wide variety of functions. Sugar and salt have been used as preservatives for centuries. Because many foods are not amenable to preservation with large concentrations of sugar or salt, other substances have been utilized. Chemical compounds and elements used as preservatives may be divided for convenience into inorganic and organic substances. The inorganic preservatives include inorganic acids and their salts, alkalis and alkaline salts, metals, halogens, peroxides and gases. The organic preservatives include organic acids and their salts, formaldehyde, sugars, alcohols, antibiotics, wood smoke, spices and condiments. Most of the better-known antibiotics have been tested on raw foods, chiefly proteinaceous ones such as poultry, meat and fish. The antibiotic nisin is employed in Europe to suppress anaerobes in cheese and cheese products.

Nisin was successfully incorporated into a calcium alginate matrix and ground into micro-particles smaller than 150 μm. Formation of micro-particles and incorporation of nisin was verified by scanning electron microscopy (SEM) and by reduction in the inactivation of nisin activity with proteolytic enzymes. Incorporation efficiency was 87 to 93% and the nisin in the alginate-incorporated form was 100% active against an indicator culture of *Lactobacillus curvatus* both in MRS broth (i.e. De Man, Rogosa and Sharpe Medium for *Lactobacilli*) and in reconstituted skim milk (Wan *et al.*, 1997). The bacteriocin divercin is active against *Listeria*, which can be produced from cells of *Carnobacterium divergens* V41 by continuous cultivation, immobilization in calcium alginate beads or in a membrane bioreactor. Immobilized cells presented the best performance (Bhugaloo Vial *et al.*, 1997). Sterilized, lean and adipose beef carcass tissues were inoculated with *Brochothrix thermosphacta*, left untreated, or treated with 100 μg ml^{-1} nisin, calcium alginate, or 100 μg ml^{-1} nisin immobilized in calcium alginate gel. The tissues were ground aseptically and nisin activity and bacterial populations of *B. thermosphacta* were determined for up to 14 days at 4°C. While no effective suppression of *B. thermosphacta* was observed, the nisin application seemed to provide some protection against other undesirable bacteria during grinding (Cutter & Siragusa, 1997).

6.5.11 Enzymes

Enzymes function as catalysts to speed up chemical reactions. Most enzymes are specific in their reactivity. Enzymes in cells remain active even when the tissue in which they are contained is dead. Enzymes can be immobilized within matrices for their targeting to specific sites or media, and to facilitate continuous, multi-functional operation. The enzyme β-galactosidase was immobilized in cobalt alginate beads. Cobalt was used as the cross-linking agent due to the higher affinity of alginate for Co^{2+} than for Ca^{2+}, a fact that caused the mechanical strength of the alginate gel to increase with increasing affinity for the divalent cation, in addition to the increase in the stability of β-galactosidase. The highest relative activity reported in the literature, 83%, was observed. Enzyme leakage was avoided by treating the surface area with glutaraldehyde. The method can be applied to other enzymes if they are not activated by cobalt salts (Ates & Mehmetoglu, 1997). Other manuscripts have described different enzymes including: conditions for pectinase production by immobilized cells of *A. niger* (Lin *et al.*, 1997); stabilization studies of L-aminoacylase-producing *Pseudomonas* sp. BA2 immobilized in a calcium alginate gel (Bodalo *et al.*, 1997); production, purification and immobilization of glucose isomerase from *Streptomyces olivochromogenes* (Azin *et al.*, 1997).

6.5.12 Water denitrification

Environmental nitrate contamination is fast becoming an international concern. Wastewater from the food and agricultural industries could be one of its major sources in our growing society. Bacterial denitrification is the most common method of biological nitrate removal. Compared to physical and chemical removal techniques, it is relatively inexpensive and reliable. Denitrification methods generally employ immobilized biosystems, i.e. physical or physico chemical bonding of denitrifiers to the surface of insoluble carriers such as sand, plastic or ceramic particles. Adsorbed microorganisms, immobilized by weak hydrogen bonds or by electrostatic interactions with the carrier (fixed film processes), can be washed easily from the support into the treated water, resulting in microbial pollution. Although enzyme entrapment, an alternative method for immobilization, has been used continuously since the 1960s, microorganism containment is a recent approach to wastewater treatment. The denitrifier, *Pseudomonas stutzeri*, entrapped in chitosan beads, was incubated under denitrifying conditions in a column receiving a continuous supply of full-strength growth medium. Biochemical, structural, and mechanical properties of the beads were studied during the first 11 days of incubation; nitrate removal was followed for 45 days. Under these conditions they contracted and became lighter with time. Contraction was not uniform in all planes. The mechanical compressive properties of the beads strengthened over the first 9 days and weakened thereafter. Concomitantly, structural changes were observed in the beads: denitrification and nitrite accumulation were lower in entrapped versus free cells (Nussinovitch *et al.*, 1996). Alginate beads containing a denitrifying isolate (*Pseudomonas* sp.) and starch were pre-treated by freeze-drying before incubation under denitrifying conditions.

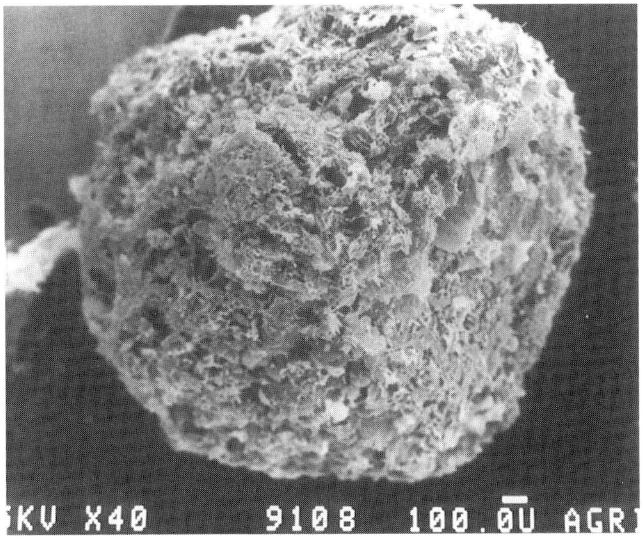

Fig. 6.6 Dried gum beads revealed physical properties similar to those of porous, sponge-like matrices. (Courtesy of Dr. Y. Tal.)

The physical and denitrifying properties of this immobilization complex were compared to those of untreated *wet* alginate beads. Freeze-dried alginate beads revealed physical properties similar to those of porous, sponge-like matrices (Fig. 6.6). As compared to conventionally prepared alginate beads, physical damage, due to gas accumulation (Fig. 6.7) under denitrifying conditions, was considerably reduced in the freeze-dried beads. Stress–strain tests were performed, and the latter beads were found to be considerably stronger than the conventionally prepared ones. The stronger, freeze-dried beads sustained denitrifying activity over a prolonged period

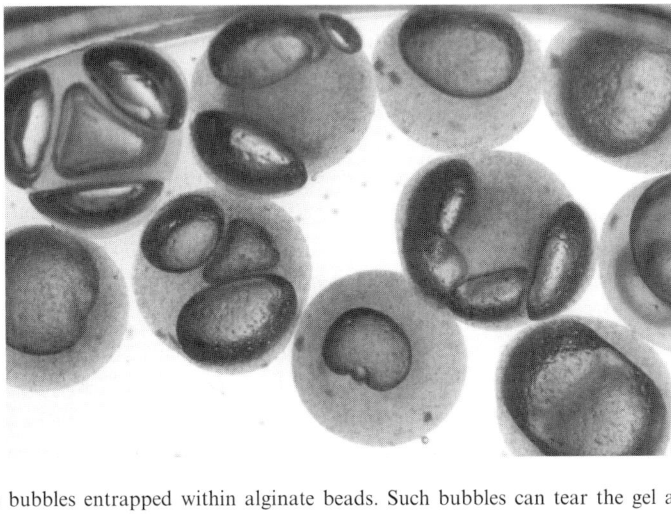

Fig. 6.7 Gas bubbles entrapped within alginate beads. Such bubbles can tear the gel and decrease its mechanical stability. (Courtesy of Dr. Y. Tal.)

relative to the regular beads (Tal *et al.*, 1997). Alginate-based beads including a denitrifying bacterium (*Pseudomonas* sp.) were freeze-dehydrated for better performance. Freeze-dried beads containing 40% granular starch had better mechanical and denitrifying properties than beads containing lower concentrations of this carbon source and filler. The biological activity, i.e. nitrate removal, was correlated with the starch content (Tal *et al.*, 1999).

Another report reported the removal of nitrate-N and organic pollutants from dairy industry wastewater by denitrification (Zayed & Winter, 1998). Artificially prepared wastewater, containing 250 mg l^{-1} nitrate-N and 1.5 g l^{-1} whey powder, was completely denitrified with removal of 90–93% of the chemical oxygen demand (COD) of the whey powder by suspended or immobilized mixed cultures, and by a suspended or immobilized pure culture that was isolated from the mixed culture inoculum. For the above COD : nitrate-N ratio of 6:1, the results indicated that the wastewater organic compounds serve as electron donors for complete denitrification and that there is no need to add an external carbon source. In batch denitrification assays, the suspended or immobilized mixed cultures proved to be more active and react faster than the isolated pure cultures. In continuous denitrification processes with immobilized pure or mixed cultures, the alginate beads, used for immobilization, were not stable for more than 12 days of incubation. The mixed free cultures removed the nitrate-N and COD continuously with no change in their activity for at least 15 days at an optimum hydraulic retention time of 0.27 days with a loading rate of 900 mg nitrate-N l^{-1} day^{-1} (van Rijn *et al.*, 2001; Zayed & Winter, 1998).

6.5.13 Aroma compounds and changes

Limonin can be effectively degraded by *Rhodococcus fascians* cells for debittering purposes. These bacteria can be entrapped in κ-carrageenan and used in a continuously stirred tank reactor to degrade limonin in a continuous process. The effects of temperature, limonin concentration, dilution rate, and aeration on reactor behavior were tested, and the results correlated with changes in limonin conversion, substrate degradation rate, and free and immobilized biomass. The immobilized cells were able to debitter limonin-containing media and the immobilized biomass was quite stable throughout the operational conditions tested (Iborra *et al.*, 1994). A population of free biomass was present in the reactor, the quantity of which was dependent on dilution rate. The immobilized bacterium increased its limonin-degrading capability when the substrate concentration was increased. The aeration was not strictly necessary for limonin degradation. Additionally, the immobilized cells were active and stable for more than 2 months of continuous operation, and were able to recover their limonin-degrading ability when used intermittently. Finally, none of the main components of a juice were noticeably altered during limonin degradation, so the reactor response was good enough to consider its application (Iborra *et al.*, 1994).

An alternative approach to the commonly practiced microbial production of bioflavors, eliminating the need for lengthy purification, was presented by Kogan and Freeman (1994). It is based on co-immobilization of precursors for bioflavor generation by microbial cells, traditionally employed for food and beverage processing, within beads made of food-grade gel matrix. Following incubation under controlled conditions, the bioflavor or bioflavor mixture is generated and accumulated within the beads. The flavor-retaining bead may then be employed as a food additive (Kogan & Freeman, 1994). Complex bioflavor generation was also demonstrated by baker's yeast co-immobilized with apple juice, generating cider flavors. Beads providing beer taste were also readily made via co-immobilization of commercial brewer's yeast with malt. Furthermore, the potential inherent in bioflavor generation by co-immobilization of filamentous fungi with an emulsion of oily precursor was demonstrated by γ-decalactone production from castor oil (Kogan & Freeman, 1994).

Biotransformation of β-ionone into hydroxy- and oxo-derivatives by *A. niger* IFO 8541 has been reported. β-Ionone, $C_{13}H_{20}O$, is a pale yellow to yellow oily liquid. It has a particularly intense freesia character, emphasized by a fruity, somewhat raspberry-like background. This product is used in all domains of perfumery and has good substantivity. Since the fungus *A. niger* develops in the form of pellets when cultivated, its entrapment within alginate beads is a natural way to mimic this feature. In the presence of a carbon source, recovery of $\sim$2.5 g l^{-1} aroma compounds after 230 h cultivation with a molar yield of $\sim$100% is possible (Larroche *et al.*, 1995). Sodium alginate beads were used for immobilization of naringinase to yield naringin hydrolysis. This can be used for debittering kinnow juice of varying pHs and temperatures (Puri *et al.*, 1996). Another interesting application is the preparation of a low-oxalate dietary preparation for hyperoxaluric patients; the high oxalate contents of spinach, amaranthus and *paruppu keerai* can be depleted by banana oxalate oxidase entrapped in alginate (Lathika *et al.*, 1995). *L. lactis* subsp. *lactis* entrapped in alginate beads was used as a biocatalyst in continuous fermentation. Activity inside the beads was limited by the reaction and internal transfer. The higher the quantity of citrate within the medium, the greater its bioconversion into aroma compounds (Cachon *et al.*, 1995b).

6.6 Factors related to bead manufacturing

Many industrial and semi-industrial processes to produce large masses of beads have been described. A rotating nozzle-ring was used to spray the alginate with the cells into a cross-linking solution. An alternative approach was to use a dual-fluid atomizer to shear sodium alginate beads off the tip of hypodermic needles by air stream into a calcium chloride solution. Many reports on the use of this popular entrapment method can be found in the literature (Grizeau & Navarro, 1986; Grote *et al.*, 1980; Klein & Kressdorf, 1983; Margaritis *et al.*, 1981; Nussinovitch *et al.*, 1994). A new process to produce beads of uniform size using sound waves has been proposed. Experiments were conducted to optimally control bead production ranges. Beads with diameters of 1.5–3.5 mm were successfully produced with low apparatus cost and easier control of shape and size (Lee *et al.*, 1996).

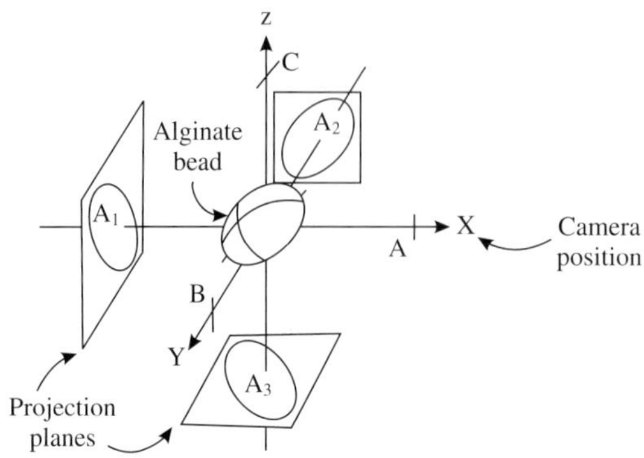

Fig. 6.8 Camera set up for recording the criterion area of hydrocolloid beads (in picture alginate bead), from which the deviations from perfect sphericity can be estimated. (After Nussinovitch & Gershon, 1996.)

6.7 Bead size, sphericity and pore-size measurements of hydrocolloid beads

The effective diffusivity of oxygen inside immobilized cell particles has been much discussed. Most reported estimates are based on fitting a mass-transfer reaction model to measured total oxygen uptake rates. The particle diameter has the largest single influence in such models, but its accurate measurement has probably received insufficient attention. Sorbitol and glucose oxidation by cells of *Gluconobacter suboxydans* entrapped in calcium alginate gel beads has been studied (Doherty *et al.*, 1995). Because these beads shrink rapidly in air, size measurement under water was essential. By comparison with rigid particles of similar known size, it was shown that measurement of the microscopic image gives a systematic underestimate. Consequently, the fitted oxygen diffusivity is around 20% too low. Careful attention to size measurement gave good agreement between diffusivity estimates from beads with different mean sizes and cell loadings, with a best value of $2.51 \times 10^9\,m^2\,s^{-1}$, 92% of the value for pure water. The estimated diffusivity is not significantly affected by a distribution of bead sizes with up to 10% standard deviation about the same mean (Doherty *et al.*, 1995). Estimation of deviations of hydrocolloid beads from sphericity is important in the study of mass and heat transfer. The sphericity of alginate beads was estimated by two techniques: a method originally developed for quartz grains expresses the shape character of the bead relative to that of the sphere of the same volume and a method previously developed for convex bodies calculates a dimensionless constant which relates surface area to volume (Fig. 6.8). Deviations from sphericity were estimated easily by the former. The latter, although more laborious, provides measurements that influence the results less. The difference between the two procedures for estimating bead sphericity was not more than 5%. Deviations from perfect sphericity are ~7% for tested alginate beads under examined conditions (Nussinovitch & Gershon, 1996).

Pores within a bead matrix are very important when size-exclusion chromatography or some other applications are being considered. Pore diameters of 0.8×10^{-8} to 1×10^{-8} m have been detected for 2% calcium alginate beads. Whey proteins such as β-lactoglobulin and α-lactoalbumin both easily penetrate such pores. In this work, no proteins had complete access to the matrix's total internal volume (Stewart & Swaisgood, 1993). The time for full gelification of sodium alginate can be estimated by using a diffusional model that considers bead diameter, temperature, concentrations of gum and cross-linking agents and inclusion of immobilized preparations within. A connection between time to complete gelation and stress at bead failure was demonstrated. The authors claimed that the presence of microorganisms increased gelling time in relation to their occupied volumes (Gilson *et al.*, 1990). Diffusion of serum albumin out of beads is dependent in part on the aforementioned factors. In addition, the importance of homogeneity was stressed as was the dependence of diffusion on temperature (Martinsen *et al.*, 1992). Possible interactions between the medium in which the beads are immersed and the gel are in part responsible for cultivation success (Quiros *et al.*, 1996).

6.8 References

Abbi, M., K.R. Chander & A. Singh (1996) Bioconversion of pentose sugars to ethanol by free and immobilized cells of *Candida shehatae* (NCL-3501): fermentation behavior. *Process Biochem.*, **31**(6), 555–560.

Abelyan, V.A. & L.A. Abelyan (1996) Production of lactic acid by immobilized cells in stirred reactors. *Appl. Biochem. Microbiol.*, **32**(5), 495–499.

Arica, M.Y., N.G. Alaeddinoglu & V. Hasirci (1998) Immobilization of glucoamylase onto activated pHEMA/EGDMA microspheres: properties and application to a packed bed bioreactor. *Enzyme and Microbial Technology*, **22**(3), 152–157.

Arnaud, J.P., C. Lacroix & F. Castaigne (1992) Counter diffusion of lactose and lactic acid in κ-carrageenan/locust bean gum gel beads with or without entrapped lactic acid bacteria. *Enzyme and Microbial Technology*, **14**(9), 715–724.

Ates, S. & U. Mehmetoglu (1997) A new method for immobilization of β-galactosidase and its utilization in a plug flow reactor. *Process Biochem.*, **32**(5), 433–436.

Azin, M., N. Moazami & M. Nemat Gorgani (1997) Production, purification and immobilization of glucose isomerase from *Streptomyces olivochromogenes* PTCC 1547. *World J. Microbiol. Biotechnol.*, **13**(5), 597–598.

Bakoyianis, V., A.A. Koutinas, K. Agelopoulos & M. Kanellaki (1997) Comparative study of kissiris, gama-alumina and calcium alginate as supports of cells for batch and continuous wine-making at low temperatures. *J. Agric. and Food Chem.*, **45**(12), 4884–4888.

Banerjee, M., A. Chakrabarty & S.K. Majumdor (1982) Immobilization of yeast cells combining β-galactosidase. *Biotechnol. Bioeng.*, **24**, 1839–1850.

Bang, W.G., U. Behrendt, S. Lang *et al.* (1983) Continuous production of L-tryptophan from indole and L-serine by immobilized *Escherichia coli* cells. *Biotechnol. Bioeng.*, **25**, 1013–1025.

Bare, G., J. Gerard, P. Jacques, V. Delaunois & P. Thonart (1992) Bioconversion of vanillin into vanillic acid by *Pseudomonas fluorescens* strain BTP9. Cell reactors and mutants study. *Appl. Biochem. Biotechnol.*, **34/35**, 499–510.

Bhugaloo Vial, P., W. Grajek, X. Dousset & P. Boyaval (1997) Continuous bacteriocin production with high cell density bioreactors. *Enzyme and Microbial Technology*, **21**(6), 450–457.

Birnbaum, S., R. Pendleton, P. Larsson *et al.* (1981) Covalent stabilisation of alginate gel for entrapment of living whole cells. *Biotechnol. Lett.*, **3**, 393–400.

Bodalo, A., J. Bastida, J.L. Gomez, I. Alcaraz & M.L. Asanza (1997) Stabilization studies of L-aminoacylase producing *Pseudomonas* sp. BA2 immobilized in calcium alginate gel. *Enzyme and Microbial Technology*, **21**(1), 64–69.

Brenna, O. & E. Bianchi (1994) Immobilized laccase for phenolic removal in must and wine. *Biotechnol. Lett.*, **16**(1), 35–40.

Brodelius, P. & K. Nilsson (1980) Entrapment of plant cells in different matrices. *Federation of European Biochemical Societies Lett.*, **122**, 312–316.

Buchholz, K., M. Noll-Borchers & D. Schwengers (1998) Production of leucrose by dextransucrase. *Starch*, **50**(4), 164–172.

Bucke, C. (1983) Immobilized cells. *Philosophical Transactions of the Royal Society*. London, Series B, **300**, 369–389.

Buzzini, P. & J. Rossi (1998) Semi-continuous and continuous riboflavin production by calcium-alginate immobilized *Candida tropicalis* in concentrated rectified grape must. *World J. Microbiol. Biotechnol.*, **14**(3), 377–381.

Cachon, R., M. Catte, R. Nomme, H. Prevost & C. Divies (1995a) Kinetic behavior of *Lactococus lactis* ssp. *lactis* bv. *diacetylactis* immobilized in calcium alginate gel beads. *Process Biochem.*, **30**(6), 503–510.

Cachon, R., P. Molin & C. Divies (1995b) Modeling of continuous pH-stat stirred tank reactor with *Lactococcus lactis* ssp. *lactis* bv. *diacetylactis* immobilized in calcium alginate gel beads. *Biotechnol. Bioeng.*, **47**(5), 567–574.

Carrara, C.R. & A.C. Rubiolo (1997) A method for evaluating lactose hydrolysis in a fixed bed reactor with β-galactosidase immobilized on chitosan. *Chemical Eng. J.*, **65**(2), 93–98.

Champagne, C.P., C. Gaudy, D. Poncelet & R.J. Neufeld (1992) *Lactococcus lactis* release from calcium alginate beads. *Appl. Environ. Microbiol.*, **58**(5), 1429–1434.

Cheetham, P.S.J. (1980) Developments in the immobilization of microbial cells and their applications, In: (Wiseman, A., Ed.), *Enzyme and Fermentation Biotechnology*. Ellis Horwood Ltd., Chichester, **4**, pp. 189–238.

Chen, K.C., S.J. Chen & J.Y. Houng (1996) Improvement of gas permeability of denitrifying PVA gel beads. *Enzyme and Microbial Technology*, **18**(7), 502–506.

Chibata, I. (1981) Immobilized microbial cells with polyacrylamide gel and carrageenan and their industrial applications. In: (Venkatsubramanian, K., Ed.), *Immobilized Microbial Cells Am. Chem. Soc. Symp.* Ser., **106**, 187–202.

Chih, Yu C., J.D. Kow, C.S. Dey, L. Chi Tsai & Y.L. Shin (1996) Production of fructooligosaccharides by immobilized mycelium of *Aspergillus japonicus*. *J. Chem. Technol. Biotechnol.*, **66**(2), 135–138.

Cruz, R., V.D. Cruz, M.Z. Belini, J.G. Belote & C.R. Vieira (1998) Production of fructo-oligosaccharides by the mycelia of *Aspergillus japonicus* immobilized in calcium alginate. *Bioresour. Technol.*, **65**(1/2), 139–143.

Cutter, C.N. & G.R. Siragusa (1997) Growth of *Brochothrix thermosphacta* in ground beef following treatments with nisin in calcium alginate gels. *Food Microbiol.*, **14**(5), 425–430.

Dainty, A.L., K.H. Goulding, P.K. Robinson, I. Simpkins & M.D. Trevan (1986) Stability of alginate-immobilized algal cells. *Biotechnol. Bioeng.*, **28**, 210–216.

Dallmann, K., L. Orosz & B. Szajani (1997) Prolonged production of tryptophan using immobilized bacteria. *Biotechnol. Lett.*, **19**, 123–125.

Davidson, R.L. (1980) *Handbook of Water-Soluble Gums and Resins*. McGraw-Hill Book Company, New York.

Deo, Y.M. & G.M. Gaucher (1983) Semi-continuous production of the antibiotic patulin by immobilized cells of Pennicillium urticae. *Biotechnol. Lett.*, **5**, 125–130.

Deriso, L., E. De-Alteriis, F. La-Cara, A. Sada & P. Parascandola (1996) Immobilization of Bacillus-Acidocaldarius whole-cell rhodanese in polysaccharide and insolubilized gelatin gels. *Biotechnol. Appl. Biochem.*, **23**, 127–131.

Dinelli, D. (1972) Entrapment in solid fibers. *Process Biochem.*, **7**(8), 9–12.

Dobreva, E., V. Ivanova, A. Tonkova & E. Radulova (1996) Influence of the immobilization conditions on the efficiency of α-amylase production by Bacilus-Licheniformis. *Process Biochem.*, **31**(3), 229–234.

Doherty, S.E., P.J. Halling & B. McNeil (1995) The importance of bead size measurement in mass-transfer modelling with immobilized cells. *Appl. Microbiol. Biotechnol.*, **43**(3), 440–444.

Dominguez, J.M. (1998) Xylitol production by free and immobilized *Debaryomyces hansenii*. *Biotechnol. Lett.*, **20**(1), 53–56.

Dong, G.Q., R. Kaul & B. Mattiasson (1991) Evaluation of alginate-immobilized *Lactobacillus casei* for lactate production. *Appl. Microbiol. Biotechnol.*, **36**(3), 309–314.

Doores, S. (1990) pH control agents and acidulants. In: (Lary Bransen, A., P. Michael Davidson & S. Salminen, Eds.), *Food Addit*. Marcel Dekker, Inc., NY and Basel, pp. 477–511.

Eskin, M.N.A. (1990) *Biochemistry of Foods*. Academic Press, San Diego, CA.

Figueiredo, Z.M.B. & L.B. Carvalho Jr. (1991) L-Malic acid production using immobilized *Saccharomyces cerevisiae*. *Appl. Biochem. Biotechnol.*, **30**(2), 217–224.

Fumi, M.D., A. Silva, G. Battistotti & O. Colagrande (1992) Living immobilized acetobacter in Ca-alginate in vinegar production: preliminary study on optimum conditions for immobilization. *Biotechnol. Lett.*, **14**(7), 605–608.

Gates, J.C. (1981) *Basic Foods*. Holt, Rinehart and Winston, New York.

Gemeiner, P., L. Kurillova, O. Markovic, A. Malovikova, D. Uhrin, M. Ilavsky, V. Stefuca, M. Polakovic & V. Bales (1991) Calcium pectate gel beads for cell entrapment. III. Physical properties of calcium pectate and calcium alginate gel beads. *Biotechnol. Appl. Biochem.*, Orlando, FL, Academic Press. June 1991. **13**, 335–345.

Gilson, C.D., A. Thomas & F.R. Hawkes (1990) Gelling mechanisms of alginate beads with and without immobilized yeast. *Process Biochemistry International*, **25**(3), 104–108.

Green, K.D., I.S. Gill, J.A. Khan *et al.* (1996) Microencapsulation of yeast cells and their use as a biocatalyst in organic solvents. *Biotechnol. Bioeng.*, **49**(5), 535–543.

Grizeau, D. & J.M. Navarro (1986) Glycerol production by *Dunaliella tertiolecta* immobilized with Ca alginate beads. *Biotechnol. Lett.*, **8**, 261–264.

Groboillot, A.F., C.P. Champagne, G.D. Darling, D. Poncelet & R.J. Neufeld (1993) Membrane formation by interfacial cross-linking of chitosan for microencapsulation of *Lactococcus lactis*. *Biotechnol. Bioeng.*, **42**(10), 1157–1163.

Grote, W., K.J. Lee & P.L. Rogers (1980) Continuous ethanol production by immobilized cells of *Zymomonas mobilis*. *Biotechnol. Lett.*, **2**, 481–486.

Gu, Z. (1997) Process development of propionic acid production by fermentation. *Dissertation Abstracts International*, B, **58**(6), 3177 Order no. DA9737713.

Guardiola, J., J.L. Iborra, L. Rodenas & M. Canovas (1996) Biotransformation from Geraniol to Nerol by immobilized Grapevine cells (*V-Vinifera*). *Appl. Biochem. Biotechnol.*, **56**(2), 169–180.

Hamamci, H. & D.D.Y. Ryu (1994) Production of L(+)-lactic acid using immobilized *Rhizopus oryzae*. Reactor performance based on kinetic model and simulation. *Appl. Biochem. Biotechnol.*, **44**(2), 125–133.

Hamdy, O.S., S. Honecker & H.J. Rehm (1992) A comparative study on the formation of citric acid and polyols and on morphological changes of three strains of free and immobilized *Aspergillus niger*. *Appl. Microbiol. Biotechnol.*, **36**(4), 518–524.

Hannoun, B.J.M. & G. Stephanopoulos (1986) Diffusion coefficients of glucose and ethanol in cell-free and cell-occupied calcium alginate membranes. *Biotechnol. Bioeng.*, **28**, 829–835.

Hoshino, K., M. Taniguchi, H. Marumoto, K. Shimizu & M. Fujii (1991) Continuous lactic acid production from raw starch in a fermentation system using a reversibility soluble-autoprecipitation amylase and immobilized cells of *Lactobacillus casei*. *Agric. Biol. Chem.*, **55**(2), 479–485.

Iborra, J.L., A. Manjon, M. Canovas, P. Lozano & C. Martinez (1994) Continuous limonin degradation by immobilized *Rhodococcus fascians* cells in κ-carrageenan. *Appl. Microbiol. Biotechnol.*, **41**(4), 487–493.

Jankovsky, M. & L. Vasakova (1996) Immobilization in alginate gels. *Veterinari Medicina*, **41**(5), 159–164.

Jen, A.C., M.C. Wake & A.G. Mikos (1996) Review – hydrogels for cell immobilization. *Biotechnol. Bioeng.*, **50**(4), 357–364.

Jiang, B., Z. Wang & X.L. Ding (1996) Production of high purity fructooligosaccharides with co-immobilized *Aspergillus niger* and enzyme. *Food & Fermentation Industries*, **1**, 1–7 (in Chinese).

Kaufman, E.N., S.P. Cooper, M.K. Budner & G.R. Richardson (1996) Continuous and simultaneous fermentation and recovery of lactic acid in a biparticle fluidized-bed bioreactor. *Appl. Biochem. Biotechnol.*, **57/58**, 503–515.

Kautola, H., W. Rymowicz, Y.Y. Linko & P. Linko (1991) Production of citric acid with immobilized *Yarrowia lipolytica*. *Appl. Microbiol. Biotechnol.*, **35**(4), 447–449.

Khachatourians, G.G., J.D. Brosseau & J.J. Child (1982) Thymidine phosphorylase activity of anucleate minicells of *E. coli* immobilized in an agarose gel matrix. *Biotechnol. Lett.*, **4**, 735–740.

Khare, S.K., K. Jha & A.P. Gandhi (1994a) Use of agarose-entrapped *Aspergillus niger* cells for the production of citric acid from soy whey. *Appl. Microbiol. Biotechnol.*, **41**(5), 571–573.

Khare, S.K., J. Krishna & M.N. Gupta (1994b) Entrapment of wheat phytase in polyacrylamide gel and its application in soymilk phytate hydrolysis. *Biotechnol. Appl. Biochem.*, **19**(2), 193–198.

Kim, D.M. & H.S. Kim (1992) Continuous production of gluconic acid and sorbitol from Jerusalem artichoke and glucose using an oxidoreductase of *Zymomonas mobilis* and inulinase. *Biotechnol. Bioeng.*, **39**(3), 336–342.

Kim, M.N., F. Ergan, P. Dhulster *et al.* (1982) Steroid modification with immobilized mycelium of *Aspergillus phoenics*. *Biotechnol. Lett.*, **4**, 233–238.

Klein, J. & B. Kressdorf (1983) Improvement of productivity and efficiency in ethanol production with Ca-alginate immobilized *Zymomonas mobilis*. *Biotechnol. Lett.*, **5**, 497–502.

Klein, J., U. Hackel, P. Schara, P. Washausen *et al.* (1978) Polymer entrapment of microbial cells: preparation and reactivity of catalytic systems. In: (Broun, G.B., G. Manecke & L.M. Wingard, Eds.), *Enzyme Engineering*. Plenum Press, New York, **4**, pp. 339–341.

Kluge, M., J. Klein & F. Wagner (1982) Production of 6-aminopenicillanic acid by immobilized *Pleurotus ostreatus*. *Biotechnol. Lett.*, **4**, 293–296.

Knorr, D., Y. Pandya & H. Doernenburg (1990) Immobilization of plant cell culture in complex coacervate capsules for secondary metabolite production. *Food Biotechnol.*, **4**(1), 409–420.

Kogan, N. & A. Freeman (1994) Development of macrocapsules containing bioflavors generated in situ by immobilized cells. *Process Biochem.*, **29**(8), 671–677.

Krisch, J. & B. Szajani (1996) Effects of immobilization on biomass production and acetic acid fermentation of *Acetobacter aceti* as a function of temperature and pH. *Biotechnol. Lett.*, **18**(4), 393–396.

Krouwel, P.G., A. Harder & N.W.F. Kossen (1982) Tensile stress-strain measurements of materials used for immobilization. *Biotechnol. Lett.*, **4**, 103–108.

Kuriki, T., M. Yanase, H. Takata & S. Okada (1997) Production of isomalto/branched oligosaccharide syrup by using immobilized neopullulanase and preliminary evaluation of the syrup as a food additive. *J. Appl. Glycoscience*, **44**(1), 15–22.

Kurillova, L., P. Gemeiner, M. Ilavsky, V. Stefuca, M. Polakovic, A. Welwardova & D. Toth (1992) Calcium pectate gel beads for cell entrapment. IV. Properties of stabilized and hardened calcium pectate gel beads with and without cells. *Biotechnol. Appl. Biochem.*, **16**(3), 236–251.

Kwak, M.Y. & J.S. Rhee (1992) Cultivation characteristics of immobilized *Aspergillus oryzae* for kojic acid production. *Biotechnol. Bioeng.*, **39**(9), 903–906.

Lacroix, C., C. Paquin & J.P. Arnaud (1990) Batch fermentation with entrapped growing cells of *Lactobacillus casei*. Optimization of the rheological properties of the entrapment gel matrix. *Appl. Microbiol. Biotechnol.*, **32**(4), 403–408.

Lamboley, L., C. Lacroix, C.P. Champagne & J.C. Vuillemard (1997) Continuous mixed strain mesophilic lactic starter production in supplemented whey permeate medium using immobilized cell technology. *Biotechnol. Bioeng.*, **56**(5), 502–516.

Larroche, C., C. Creuly & J.B. Gros (1995) Fed-batch biotransformation of β-ionone by *Aspergillus niger*. *Appl. Microbiol. Biotechnol.*, **43**(2), 222–227.

Lathika, K.M., S. Sharma, K.V. Inamdar & K.G. Raghavan (1995) Oxalate depletion from leafy vegetables using alginate entrapped banana oxalate oxidase. *Biotechnol. Lett.*, **17**(4), 407–410.

Leach, G., G. Oliveira & R. Morais (1998) Production of carotenoid rich product by alginate entrapment and fluid bed drying of *Dunaliella salina*. *J. Sci. Food Agric.*, **76**(2), 298–302.

Lebeau, T., T. Jouenne & G.A. Junter (1997) Fermentation of D-xylose by free and immobilized *Saccharomyces cerevisiae*. *Biotechnol. Lett.*, **19**(7), 615–618.

Lebeau, T., T. Jouenne & G.A. Junter (1998) Diffusion of sugars and alcohols through composite membrane structures immobilizing viable yeast cells. *Enzyme Microb. Technol.*, **22**(6), 434–438.

Lee, H.H., O.J. Park, J.M. Park & J.W. Yang (1996) Continuous production of uniform calcium alginate beads by sound wave induced vibration. *J. Chem. Technol. Biotechnol.*, **67**(3), 255–259.

Li, R.H., D.H. Altreuter & F.T. Gentile (1996) Transport characterization of hydrogel matrices for cell encapsulation. *Biotechnol. Bioeng.*, **50**(4), 365–373.

Lin, Y.H., X.M. Liu, Y.F. Shen & Y.L. Chen (1997) Conditions for pectinase production by immobilized cells of *Aspergillus niger*. *Food Sci., China*, **18**(2), 16–20 (in Chinese).

Linko, M. (1974) Enzymes in the food industry. *Kemia Kemi*, **1**(10), 691–695.

Linko, Y.Y., L. Pohjola & P. Linko (1977) Entrapped glucose isomerase for high fructose syrup production. *Process Biochem.*, **12**(6), 14–16.

Margaritis, A., P.K. Bajpal & J.B. Wallace (1981) High ethanol productivity using small Ca-alginate beads of immobilized cells of Zymomonas mobilis. *Biotechnol. Lett.*, **3**, 613–618.

Martino, A., P.G. Pifferi & G. Spagna (1996) Immobilization of β-glucosidase from a commercial preparation. II. Optimization of the immobilization process on chitosan. *Process Biochem.*, **31**(3), 287–293.

Martinsen, A., G. Skjak-Break & O. Smidsrod (1989) Alginate as immobilization material. I. Correlation between chemical and physical properties of alginate gel beads. *Biotechnol. Bioeng.*, **33**(1), 79–89.

Martinsen, A., I. Storro & G. Skjak-Braek (1992) Alginate as immobilization material. III. Diffusional properties. *Biotechnol. Bioeng.*, **39**(2), 186–194.

Mattiasson, B. (1983) *Immobilized Cells and Organelles*, vols. 1 & 2. CRC Press, Boca Raton, FL.

Morin, N., C.M. Bernier & C.P. Champagne (1992) Production of concentrated *Lactococcus lactis* subsp. *cremoris* suspensions in calcium alginate beads. *Appl. Environ. Microbiol.*, **58**(2), 545–550.

Mosbach, K. & R. Mosbach (1966) Entrapment of enzymes and micro-organisms in synthetic cross-linked polymers and their application in column techniques. *Acta Chem. Scand.*, **20**, 2807–2810.

Nampoothiri, K.M. & A. Pandey (1998) Immobilization of *Brevibacterium* cells for the production of L-glutamic acid. *Bioresource Technol.*, **63**(1), 101–106.

Neufeld, R.J., Y. Peleg, J.S. Rokem, O. Pines & I. Goldberg (1991) L-malic acid formation by immobilized *Saccharomyces cerevisiae* amplified for fumarase. *Enzyme Microb. Technol.*, **13**(12), 991–996.

Norton, S., C. Lacroix & J.C. Vuillemard (1994) Kinetic study of continuous whey permeate fermentation by immobilized *Lactobacillus helveticus* for lactic acid production. *Enzyme Microb. Technol.*, **16**(6), 457–466.

Norton, S., K. Watson & T. D'Amore (1995) Ethanol tolerance of immobilized brewers' yeast cells. *Appl. Microbiol. Biotechnol.*, **43**(1), 18–24.

Nussinovitch, A. (1994) Resemblance of immobilized *Trichoderma viride* fungal spores in an alginate matrix to a composite material. *Biotechnol. Prog.*, **10**, 551–554.

Nussinovitch, A. (1997) *Hydrocolloid Applications: Gum Technology in the Food and Other Industries*. Blackie Academic & Professional, London.

Nussinovitch, A. & Z. Gershon (1996) A rapid method for determining sphericity of hydrocolloid beads. *Food Hydrocolloids*, **10**(2), 263–266.

Nussinovitch, A., M. Nussinovitch, R. Shapira & Z. Gershon (1994) Influence of immobilization of bacteria, yeasts and fungal spores on the mechanical properties of agar and alginate gels. *Food Hydrocolloids*, **8**, 361–372.

Nussinovitch, A., Y. Aboutboul, Z. Gershon & J. van Rijn (1996) Changes in mechanical, structural, and denitrifying properties of entrapped *Pseudomonas stutzeri* bacteria preparations. *Biotechnol. Prog.*, **12**(1), 26–30.

Ochi, H., M. Takahashi, T. Kaneko, H. Suzuki & H. Tanaka (1991) Diacetyl production by co-immobilized citrate positive *Lactococcus lactis* subsp. *lactis* 3022 and homogenized bovine liver in alginate fibres with double gel layers. *Biotechnol. Lett.*, **13**(7), 505–510.

Ogbonna, J.C., Y. Amano & K. Nakamura (1989) Elucidation of optimum conditions for immobilization of viable cells by using calcium alginate. *J. of Fermentation and Bioengineering*, **67**(2), 92–96.

Oliveira, E.A., A.A.R. Costa, Z.M.B. Figueiredo & L.B. Carvalho Jr. (1994) L-malic acid production by entrapped *Saccharomyces cerevisiae* into polyacrylamide gel beads. *Appl. Biochem. Biotechnol.*, **47**(1), 65–72.

Olston, A.C. & R.A. Korus (1976) Immobilized enzymes and other new applications of enzymes to food processing. *Abstracts of Papers, American Chemical Society*, **172**, AGFD 19.

Oyaas, J., I. Storro, H. Svendsen & D.W. Levine (1995) The effective diffusion coefficient and the distribution constant for small molecules in calcium-alginate gel beads. *Biotechnol. Bioeng.*, **47**(4), 492–500.

Pandya, Y. & D. Knorr (1991) Diffusion characteristics and properties of chitosan coacervate capsules. *Process Biochem.*, **26**(2), 75–81.

Parajo, J.C., H. Dominguez & J.M. Dominguez (1996) Production of xylitol from concentrated wood hydrolysates by *Debaryomyces hansenii*: Effect of the initial cell concentration. *Biotechnol. Lett.*, **18**(5), 593–598.

Pegg, R.B. & F. Shahidi (1999) Encapsulation and controlled release in food preservation. In: (Shafiur Rahman, M., Ed.), *Handbook of Food Preservation*. Marcel Dekker, Inc., New York and Basel, pp. 611–667.

Perols, C., B. Piffaut, J. Scher, J.P. Ramet & D. Poncelet (1996) The potential of enzyme entrapment in konjac cold-melting gel beads. *Enzyme and Microbial Technology*, **20**(1), 57–60.

Prevost, H. & C. Divies (1992) Cream fermentation by mixed culture of lactococci entrapped in two-layer calcium alginate beads. *Biotechnol. Lett.*, **14**(7), 583–588.

Puri, M., S.S. Marwaha & R.M. Kothari (1996) Studies on the applicability of alginate-entrapped Naringinase for the debittering of Kinnow Juice. *Enzyme and Microbial Technology*, **18**(4), 281–285.

Quiros, C., L.A. Garcia & M. Diaz (1996) The evolution of the structure of calcium alginate beads and cell leakage during protease production. *Process Biochem.*, **31**(8), 813–822.

Rehr, B., C. Wilhelm & H. Sahm (1991) Production of sorbitol and gluconic acid by permeabilized cells of *Zymomonas mobilis*. *Appl. Microbiol. Biotechnol.*, **35**(2), 144–148.

Rickert, D.A., C.E. Glatz & B.A. Glatz (1998) Improved organic acid production by calcium alginate-immobilized propionibacteria. *Enzyme and Microbial Technology*, **22**(5), 409–414.

Rochefort, W.E., T. Rehg & P.C. Chau (1986) Trivalent cation stabilization of alginate gel for cell immobilization. *Biotechnol. Lett.*, **8**, 115–120.

Roukas, T. & P. Kotzekidou (1996) Continuous production of lactic acid from deproteinized whey by coimmobilized *Lactobacillus casei* and *Lactococcus lactis* cells in a packed bed reactor. *Food Biotechnol.*, **10**(3), 231–242.

Roukas, T. & P. Kotzekidou (1998) Lactic acid production from deproteinized whey by mixed cultures of free and coimmobilized *Lactobacillus casei* and *Lactococcus lactis cells* using fedbatch culture. *Enzyme and Microbial Technology*, **22**(3), 199–204.

Ruggeri, B., G. Sassi, V. Specchia, F. Bosco & M. Marzona (1991) Alginate beads coated with polyacrylamide resin: potential as biocatalyst. *Process Biochem.*, **26**(6), 331–335.

Rymowicz, W., H. Kautola, M. Wojtatowicz, Y.Y. Linko & P. Linko (1993) Studies on citric acid production with immobilized *Yarrowia lipolytica* in repeated batch and continuous air-lift bioreactors. *Applied Microbiol. Biotechnol.*, **39**(1), 1–4.

Salminen, S. & A. Hallikainen (1990) Sweeteners. In: (Lary Bransen, A., P. Michael Davidson & S. Salminen, Eds.), *Food Additives*. Marcel Dekker, Inc., NY and Basel, pp. 297–327.

Shiow, L.L., Y.C. Hsin, C.C. Wen & C.C. Cheng (1998) Production of gamma-decalactone from ricinoleic acid by immobilized cells of *Sporidiobolus salmonicolor*. *Process Biochem.*, **33**(4), 453–459.

Shoichet, M.S., R.H. Li, M.L. White *et al.* (1996) Stability of hydrogels used in cell encapsulation-an in vitro comparison of alginate and agarose. *Biotechnol. Bioeng.*, **50**(4), 374–381.

SivaRaman, H., B.S. Rao, A.V. Pundle *et al.* (1982) Continuous ethanol production by yeast cells immobilized in open pore gelatin matrix. *Biotechnol. Lett.*, **4**, 359–364.

Smiley, K.L. (1976) Immobilized cells and enzymes for foods. *Abstracts of Papers, American Chemical Society*, **171** (centennial), MICR 11.

Smiley, K.L. & G.W. Strandberg (1972) Immobilized enzymes. *Advances in Applied Microbiology*, **15**, 13–38.

Sodini, I., C.Y. Boquien, G. Corrieu & C. Lacroix (1997) Microbial dynamics of co- and separately entrapped mixed cultures of mesophilic lactic acid bacteria during the continuous prefermentation of milk. *Enzyme and Microbial Technology*, **20**, 381–388.

Stewart, W.W. & H.E. Swaisgood (1993) Characterization of calcium alginate pore diameter by size-exclusion chromatography using protein standards. *Enzyme and Microbial Technology*, **15**(11), 922–927.

Stocklein, W., A. Eisgruber & H.L. Schmidt (1983) Conversion of L-phenylalanine to L-tyrosine by immobilized bacteria. *Biotechnol. Lett.*, **5**, 703–708.

Sutherland, J.P., A.H. Varnam & M.G. Evans (1986) *A Color Atlas of Food Quality Control*. A Wolfe Science Book, Weert, The Netherlands.

Tal, Y., J. van Rijn & A. Nussinovitch (1997) Improvement of structural and mechanical properties of denitrifying alginate beads by freeze-drying. *Biotechnol. Prog.*, **13**(6), 788–793.

Tal, Y., J. van Rijn & A. Nussinovitch (1999) Improvement of mechanical and biological properties of freeze-dried denitrifying alginate beads by using starch as a filler and carbon source. *Appl. Microbiol. Biotechnol.*, **51**(6), 773–779.

Tampion, J. & M.D. Tampion (1987) *Immobilized Cells: Principles and Applications*. Cambridge University Press, Cambridge.

Ullrich, S., U. Werrmann & D. Knorr (1990) Development of pectin–chitosan coacervate capsules for immobilizing of cultured plant cells. *Food Biotechnol.*, **4**(1), 489.

van Rijn, J., A. Nussinovitch & Y. Aboutbul (2001) Means and process for nitrate removal. United States Patent 6,297,033.

Van Vliet, T., J. Tramper & R.H. Wijffels (1997) Relevance of rheological properties of gel beads for their mechanical stability in bioreactors. *Biotechnol. Bioeng.*, **56**(5), 517–529.

Venkatasubramanian, K. & W.R. Vieth (1979) Immobilized microbial cells fermentation processes. *Prog. Ind. Microbiol.*, **15**, 61–86.

Vorlop, K.D. & J. Klein (1981) Formation of spherical chitosan biocatalysts by iontropic gelation. *Biotechnol. Lett.*, **3**, 9–14.

Walsh, P.K., F.V. Isdell, S.M. Noone *et al.* (1996a) Microcolonies in alginate and carrageenan gel particles effect of physical and chemical properties of gels. *Enzyme and Microbial Technology*, **18**(5), 366–372.

Wan, J., J.B. Gordon, K. Muirhead, M.W. Hickey & M.J. Coventry (1997) Incorporation of nisin in micro-particles of calcium alginate. *Lett. Appl. Microbiol.*, **24**(3), 153–158.

Wang, H.Y. & D.J. Hettwer (1982) Cell immobilization in κ-carrageenan with tricalcium phosphate. *Biotechnol. Bioeng.*, **24**, 1827–1838.

Wang, H.Y., S.S. Lee, Y. Takach *et al.* (1982) Maximizing microbial cell loading in immobilized cell systems. In: (Garden, E.I., Jr. Ed.), *Biotechnol. Bioeng. Symposium 12*. John Wiley, New York, pp. 136–146.

Wheatley, M.A. & C.R. Phillips (1983) The influence of internal and external diffusional limitations on the observed kinetics of immobilized whole bacteria cells with cell-associated b-glucosidase activity. *Biotechnol. Lett.*, **5**, 79–84.

Wiksstrom, P., E. Szwajcer, P. Brodelius *et al.* (1982) Formation of α-keto acids from amino acids using immobilized bacteria and algae. *Biotechnol. Lett.*, **4**, 153–158.

Yokotsuka, K., M. Yajima & T. Matsudo (1997) Production of bottle fermented sparkling wine using yeast immobilized in double-layer gel beads or strands. *Amer. J. Enology and Viticulture*, **48**(4), 471–481.

Zayed, G. & J. Winter (1995) Batch and continuous production of lactic acid from salt whey using free and immobilized cultures of lactobacilli. *Appl. Microbiol. Biotechnol.*, **44**(3/4), 362–366.

Zayed, G. & J. Winter (1998) Removal of organic pollutants and of nitrate from wastewater from the dairy industry by denitrification. *Appl. Microbiol. Biotechnol.*, **49**(4), 469–474.

Zayed, G. & A.S. Zahran (1991) Lactic acid production from salt whey using free and agar immobilized cells. *Lett. Appl. Microbiol.*, **12**(6), 241–243.

Zhong, G.U., B.A. Glatz & C.E. Glatz (1998) Effects of propionic acid on propionibacteria fermentation. *Enzyme and Microbial Technology*, **22**(1), 13–18.

Zhou, Y., E. Martins, A. Groboillot, C.P. Champagne & R.J. Neufeld (1998) Spectrophotometric quantification of lactic bacteria in alginate and control of cell release with chitosan coating. *J. Appl. Microbiol.*, **84**(3), 342–348.

Chapter 7
Texturization of Vegetative Materials

7.1 Introduction

Many processed foods contain hydrocolloids (gums), which govern the product's functionality and organoleptic acceptability (Aguilera, 1992; Aguilera & Stanley, 1986; Matz, 1984). Hydrocolloids are natural, modified natural, synthetic or biosynthetic polymers that dissolve in water and have the ability to thicken or gel aqueous systems (Nussinovitch *et al.*, 1989; Silberberg, 1989). Examples include alginate, starch, modified starches, agar, carrageenan, gelatin, xanthan, and gum arabic, as well as some proprietary formulations. Hydrocolloids are frequently used in preparing gum-based foods, significantly affecting their appearance, physical properties and shelf-life. Hydrocolloids can be selected to suit the manufacturing conditions, and specific manufacturing processes have been invented to take advantage of the functional properties of the various gums (Nussinovitch, 1993, 1997).

7.2 Agar and alginate-based texturized fruits

Formulations or techniques for processing fruit-based products can be optimized through texture measurements, thus helping to improve the organoleptic or technological qualities of the product (Dumas, 1984; Flora & Beuchat, 1979). A method to analyze changes in firmness in texturized products prepared with passion pulp was developed to aid in overcoming manufacturing constraints with such products (Mouquet, 1996). Alginate can be used to prepare products with different textures. Plain and modified insoluble fibrous alginate can be used for preparing natural fruits, vegetables and meat analogs, while propylene glycol alginate is more suitable as a flavor and color carrier, for concentrates and flavor emulsions (Anon, 1980).

Novel structured fruit products made of pulp, a wide range of hydrocolloid gels, and other additives, have been the subject of a number of patents and commercial applications (Glicksman, 1976; Lodge, 1981; Szczesniak, 1968; Tolstoguzov, 1971). These products are, in most cases, composite materials in which particulates are embedded in a polymeric gel matrix (Ring & Stainsby, 1982). Uses of sodium alginate for food thickening, production of artificial crystallized fruits and in nougat, and the manufacture of artificial fruits from sodium alginate solutions, have been previously discussed (Less, 1998; Tateo, 1985). Many patents discuss the possibility of using a combination of alginates with other hydrocolloids such as agar and carrageenan, together with fruit pulp and other traditional food additives, to create simulated fruit products (Szczesniak, 1968; Tolstoguzov, 1971; Uniliver Ltd., 1974). The available technological information on such products concentrates mainly on the methods of producing different gel systems containing pulp, sugar and acid. Since the 1970s, many alternative processes for preparing fabricated fruits have been described in patents and technical publications (Luh *et al.*, 1976; Tolstoguzov & Braudo, 1983; Wood, 1975). Calcium alginate gels (1% algin), with or without agar (1%) as a pre-setting agent, were used to form a matrix for texturized (2–8%) raspberry pulp (Fig. 7.1). Such products, with over 10% highly acidic pulp, are so weak that they

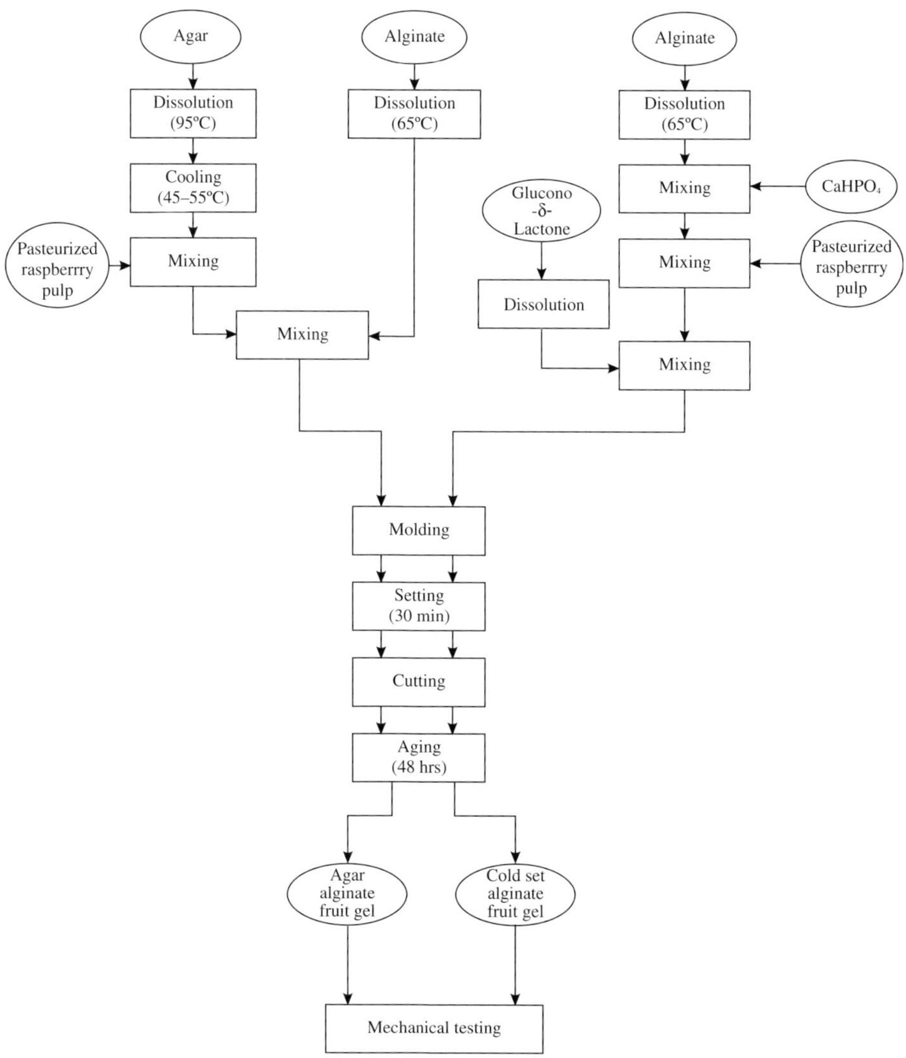

Fig. 7.1 The processes used to prepare texturized raspberry products. (From Nussinovitch & Peleg, 1990, with permission from Food & Nutrition Press, Inc.)

collapse under their own weight (Nussinovitch & Peleg, 1990). The agar did not have an appreciable effect on the product's strength, but it made it comparatively more brittle and stiffer. Although full equilibrium of such texturized products took at least 48 h or more, all of their principal mechanical features could be determined in tests performed after only 24 h (Nussinovitch & Peleg, 1990).

Alginate gels can be prepared by different methods to achieve moieties differing in their ascorbic acid retention. A combined gelatin alginate gel was created by molding the requested gel shape with gelatin and cross-linking its embedded alginate with calcium lactate, which diffused out the bath into which the combined gel was introduced. In such gels, retention of previously incorporated ascorbic acid was ~25% (the rest diffused out) after 3 days in cross-linking solution. A cold-set (internal-set) alginate preparation (using glucono-δ-lactone (GDL) to change the pH, causing a decrease in the sequestering agent's ability to entrap calcium and followed by alginate cross-linking by the liberated calcium), produced a similar gel texture but without ascorbic acid losses since no calcium diffusion bath was involved (Pelaez & Karel, 1981). Internal setting has been used to prepare homogeneous texturized fruit (Mouquet *et al.*, 1997). In some cases, the pulp is enriched with sugar before texturization to affect the texture and aroma of the product. Alginate, calcium salt and GDL concentrations have a major effect on the strength and thermostability of the formed product pieces: the higher the sweetened pulp content, the better the flavor properties of the product (Mouquet *et al.*, 1992). Texturized fruits based on passion fruit juice were produced for incorporation in bakery products, yogurts and preserves. Different sugars and sugar combinations were added to the gel mixture. The most acceptable texturized fruit was based on a sucrose/glucose syrup (Bellarde *et al.*, 1995). Sweet potato texturized puree was dependent on tetrasodium pyrophosphate, alginate and calcium sulfate contents (Truong *et al.*, 1995).

7.3 Dependence of composite fruit products on pulp properties

Pulp addition causes a reduction in the strength of texturized fruit products, down to a minimum point at a pulp concentration of about 20–30% for orange pulp (Fig. 7.2). Beyond the minimum point, the trend changes and some gel strengthening is observed (Nussinovitch *et al.*, 1991a). Stress at failure of texturized fruit depends on pulp particle size and properties. Ground pulp affects agar-based texturized fruit the least; washed pulp can restore the gel to close to its original strength. Pulp type also plays a role in determining the mechanical properties of the final texturized fruit. Banana pulp has a greater influence on the strength of texturized fruit than orange pulp. For both types of pulps, at a certain level of pulp addition (different for each pulp), the strength of the produced texturized fruit is minimum (Nussinovitch *et al.*, 1991a). The first phase in the effect of pulp on gel strength is characterized by a weakening of the gel structure through interference with matrix formation. Fine grinding minimizes this interference, indicating the importance of steric effects. At high enough pulp concentrations, if a pulp structure is formed, a reversal of the weakening trend is observed. In that case, the mechanical properties of the pulp

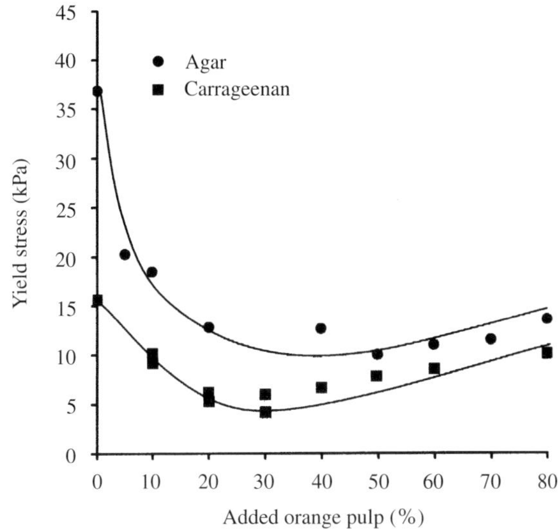

Fig. 7.2 The effect of orange pulp on the yield stress of agar and carrageenan gels. (From Nussinovitch *et al.*, 1991a.)

particles and their ability to form interacting structural elements will determine the strength of the combined gel–pulp system. Pulp-structure formation appears to depend on the shape and size of the particulates (Fig. 7.3). Ground orange pulp, lacking that structure-formation ability, causes only a decrease in gel strength with no minimum at all; nevertheless the obtained gels are stronger than those of unground pulp samples (Nussinovitch *et al.*, 1991a). During mechanical testing, if the external

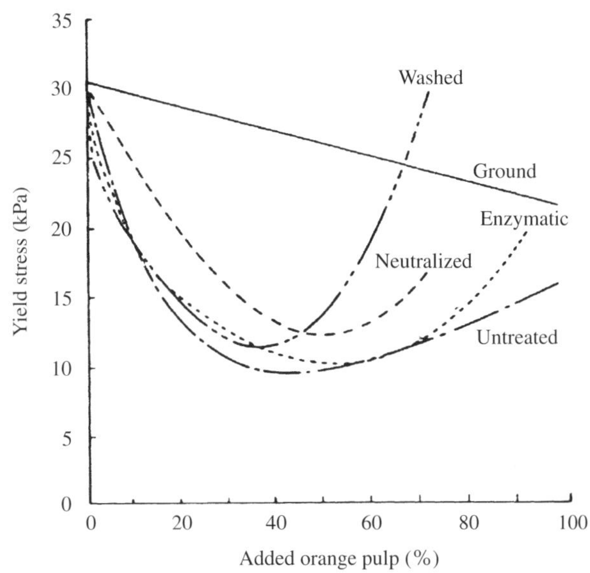

Fig. 7.3 The effect of different treatments of orange pulp on yield stress of 2% agar gel. (From Nussinovitch *et al.*, 1991a.)

pressure combined with that exerted internally by the network is higher than the osmotic pressure of the system, liquid is expelled from the gel matrix. Such liquid liberation is called *mechanical syneresis*: it decreases with increasing pulp and sugar contents. The phenomenon of minima enables the formation of texturized fruits of the same strength, which differ in their pulp content and its preceding treatment (Nussinovitch *et al.*, 1991a).

7.4 Combined effect of fruit pulp, sugar and gum on texturized fruit products

Mechanical properties of agar–sugar–pulp fruit analogs are affected by changing the concentration of any of the gel components. The gel strength and deformability modulus clearly increase with gum concentration. In principle, deformability modulus and Young's modulus are the same thing. The difference lies in the size of the deformation being addressed. Young's modulus is calculated at small deformations, whereas the deformability modulus is calculated at large deformations. Sugar produces a maximum in these two parameters, but not in the deformation at failure. The latter, however, goes through a maximum when the agar concentration is changed (Nussinovitch *et al.*, 1991b). In the GDL–alginate system, pulp produces minima in strength and deformability modulus. However, the shape of the strength versus alginate concentration curve suggests a possible maximum point (Fig. 7.4). Such a maximum is obvious in the case of deformation at failure. This is the same behavior observed in the agar gel. On the other hand, there was no observed effect of sugar on the strength or deformability modulus in GDL–alginate–sugar–pulp gels. A minimum was observed for the agar gel with respect to its pulp content. Thus, the mechanical behavior of the agar–sugar–pulp gel seems to be similar to the GDL–alginate system with regard to the effect of the pulp. A few empirical equations have described the dependence of mechanical parameters on gum, sugar and pulp concentrations. The interesting point about these equations is that a good fit is obtained in the absence of interaction terms and, thus, the empirical equations are a superposition of the effect of each component. One can obtain these equations simply by maintaining two of the independent variables constant and changing only one of them at a time. Therefore, it is possible to obtain a satisfactory quantitative estimate of the mechanical performance of these gels via a very simple procedure (Nussinovitch *et al.*, 1991b). The derivatives of these equations enable us to evaluate the maximum and minimum points. For the agar system, the maxima in yield stress and deformability modulus were obtained at the same sugar concentration, $\sim$25%, and the minimum at a pulp content of $\sim$60%. Maxima and minima clearly indicate that the mechanical properties of the agar–sugar–pulp gel are determined by counteractive processes. The pulp seems to interfere with dense gel matrix formation but contributes its mechanical properties to the structure formed. Sugar probably increases the polymer inter-molecular attraction but seems to cause gel non-homogeneity at high concentrations. The alginate system differs with respect to the minimum point obtained at $\sim$22% pulp and by the fact that no maximum is produced by varying

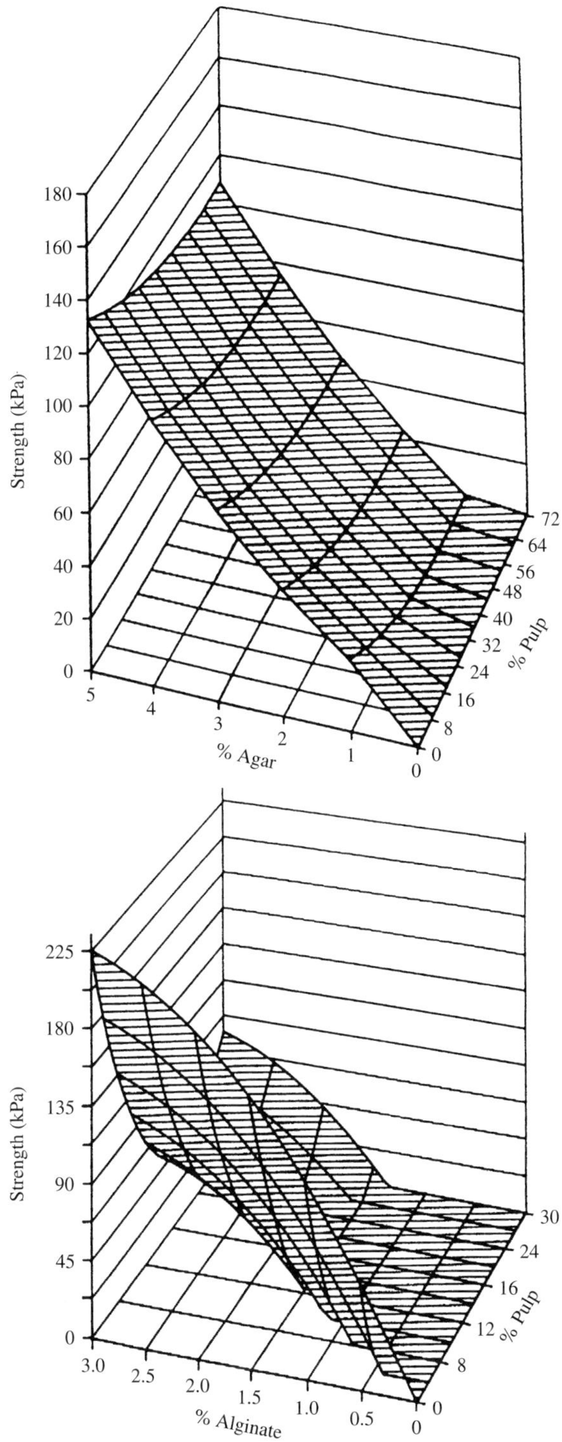

Fig. 7.4 The effect of gum and pulp concentration on the strength of agar– and alginate–sugar–pulp gels. (From Nussinovitch *et al.*, 1991b.)

the sugar concentration. Alginate concentration produces a maximum in the deformation at failure at a concentration of 1.8% and probably in the yield stress at concentrations above 3%. The presence of maxima and minima in the mechanical behavior of gum–sugar–pulp gels poses an experimental problem when the range of changes in the mechanical parameters has to be established. As already mentioned, the experimental problem can be solved by superimposing the effect of each variable on its weight. On the other hand, these maxima and minima provide invaluable technological flexibility in terms of obtaining a desired mechanical attribute by using different combinations of gum, sugar and pulp concentrations. The flexibility is limited when the degree of freedom in selecting pulp and sugar concentrations is limited or when specifying a number of desired mechanical attributes that are to be met in product formulation (Nussinovitch *et al.*, 1991b).

7.5 Alginate texturization of highly acidic fruit pulps and juices

Inclusion of highly acidic fruit pulps and juices in the composition of texturized fruit influences the latter's texture, taste and acceptability. The pulp's low pH limits its inclusion in the texturized fruit product, since it causes self-disintegration or a reduction in the product's ability to contain as many particles, both of which interfere with the product's inner structure. An interesting approach to changing pulp properties in order to enable its inclusion at high proportions within the texturized fruit was attempted by Kaletunc *et al.* (1990). Applesauce (13°Bx) and frozen grapefruit juice concentrate (37°Bx), both with low pH (~3.4–3.5), were used as raw materials. Both were neutralized with sodium hydroxide, and their total solid content was adjusted to 10°Bx. They were then added to a cold-set alginate gel (Nussinovitch, 1997). In this system, the GDL acidified the texturized fruit product to pH 4.1–4.5 for the apple and 4.3–5.0 for the grapefruit juice. This procedure paved the way for texturization of acidic ingredients by neutralization, followed by acidification once the gel is produced. When the gel product is finished, it is immersed in an acid dip to further adjust it to the original fruit ingredient's pH. Inclusion of the pulp reduced syneresis from the product, from ~49% for pure gel to 9 and 10% in the apple and grapefruit texturized products, respectively (Kaletunc *et al.*, 1990). The strength (stress at failure), strain at failure and deformability modulus of texturized apple pulp as a function of pulp concentration are shown in Fig. 7.5. These three mechanical parameters decreased considerably with increasing pulp concentration. However, even products with a considerable proportion of fruit components (50–96%) had good texture and kept their integrity, in comparison to products for which 10% untreated fruit pulp was beyond their practical limits of inclusion to obtain a reasonable product. The three aforementioned mechanical properties increased, without exception, after dipping the texturized products in a calcium lactate solution (Fig. 7.5). Dipping in dilute citric acid restored the pH to 3.2–3.4, while having only a marginal effect on the mechanical behavior of the products (Kaletunc *et al.*, 1990). With texturized grapefruit concentrate, as with texturized apple pulp, pH adjustment resulted in products with appreciable mechanical integrity, even at juice concentrations

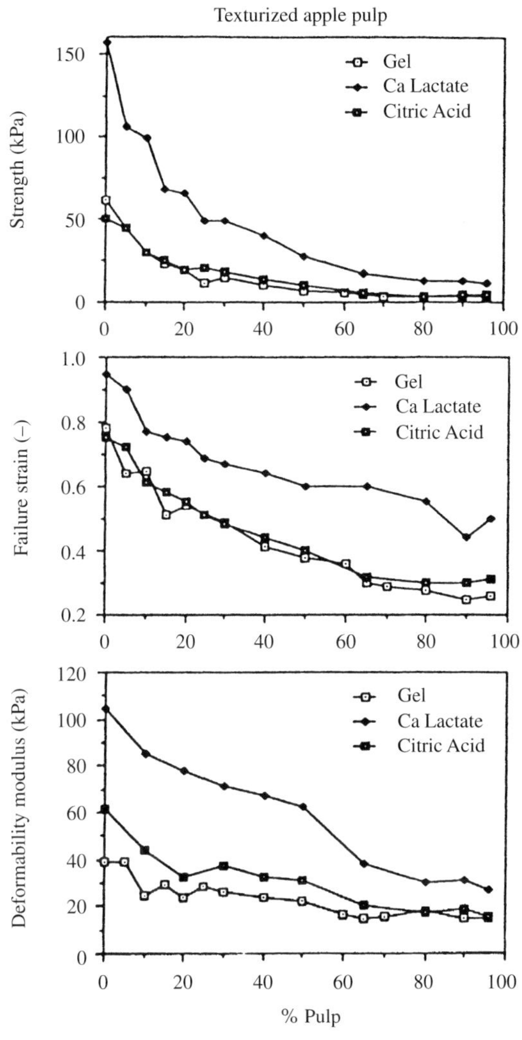

Fig. 7.5 Mechanical properties of texturized apple pulp products as a function of pulp concentration. (From Kaletunc *et al.*, 1990.)

of more than 90%. The effect of calcium lactate was also studied separately in the grapefruit products by changing exposure conditions. The gels were dipped in a mixture of calcium lactate and citric acid, or alternatively, first in an acid bath and then in a calcium lactate solution. The results are presented in Fig. 7.6. The firmness of the texture depended on reactivity with calcium ions. The order of pH adjustment, i.e. whether calcium lactate was added together with acid or after the acid treatment, did not appear to be significant (Kaletunc *et al.*, 1990).

7.6 Succulent, hydrocolloid-based, texturized fruit products

Although the technology involved in including a high proportion of fruit pulp or juice (Kaletunc *et al.*, 1990) has improved, texturized fruits still suffer from some

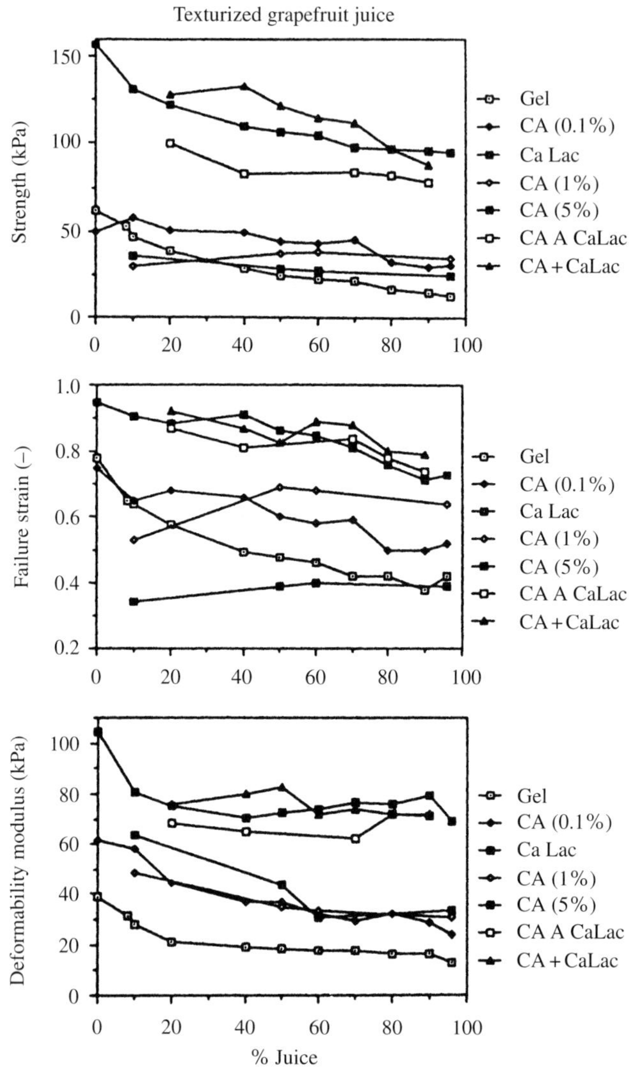

Fig. 7.6 Mechanical properties of grapefruit juice products as a function of juice concentration. CA A CaLac: dip in a calcium lactate solution and subsequently in citric acid. CA + CaLac: dip in a mixed solution of citric acid and calcium lactate. (From Kaletunc *et al.*, 1990.)

limitations. Often, their texture does not resemble that of real fruit. One of the reasons for this is that during mastication, fruit juice or liquid is only liberated to a small extent in texturized fruits, as compared to what happens when consuming fresh citrus fruits.

Pasteurized grapefruit juice cells (Fig. 7.7) were obtained by the following method. Grapefruits were washed and then passed on conveyors through a steam chamber for about 5 min, to loosen the peel from the pulp. The fruits were then cooled by cold-water spray as they emerged from the scalding tanks. The peel was removed by hand. After sectioning, the segments were passed through a $20\,g\,L^{-1}$ lye solution for approximately 30 s to remove the remaining albedo and segment membranes.

Fig. 7.7 Grapefruit vesicles following wax removal and hot-water wash. (From Weiner & Nussinovitch, 1994.)

Lye was removed from the fruit surfaces by spraying with water. After removal of the carpellary membranes and washing, isopropyl alcohol was used to remove the wax from the outer surface of the juice vesicles, with the aim of retaining as many intact cells as possible. Cells were pasteurized (82°C for 3 min) before being mixed with the agar or alginate solutions (Fig. 7.8). Hydrocolloid solutions of agar and cold-set alginate containing the pasteurized grapefruit juice cells were poured into cylindrical molds and allowed to cool and gel. Each juice vesicle, although entrapped within the matrix, did not adhere to it, but rather seemed to remain in a cavity within the gel. If the juice cell did indeed touch the walls of the cavity, the interaction between cells and matrices was very weak because juice cells could be very easily separated. Agar and alginate produced the matrices of the texturized fruits. Although they are totally different in their chemical compositions and gelation mechanisms, they behaved similarly in terms of changes in strength. In both cases, stress at failure decreased as the percentage of embedded juice sacs increased. It was not possible to build an agar matrix with more than 350 g kg^{-1} cells. With alginate, up to 400 g kg^{-1} cells could be added before a very weak, texturally unacceptable product was obtained. In another study, only 100 g kg^{-1} raspberry pulp could be embedded in a 10 g kg^{-1} alginate matrix before the product began creeping under its own weight (Nussinovitch & Peleg, 1990). Different results were achieved when commercial pulps, such as banana and orange, were used. This discrepancy is due to the fact that in such fruit pulp (depending on the degree to which it is ground), the number of particles is huge and one can find small particles of the order of a few microns next to particles measuring hundreds of microns. Juice vesicles (1–8 mm) from commercial pulp particles are more uniform, and, of course, less cells are needed to fill the volume of the matrix. The incorporation of particles, or fibers, into the matrix somewhat resembles the technique of producing a composite material with polymers, although in our case no increase in strength or stiffness was achieved, as occurs with other composite materials. This fact can be seen as indirect evidence for a weak or no

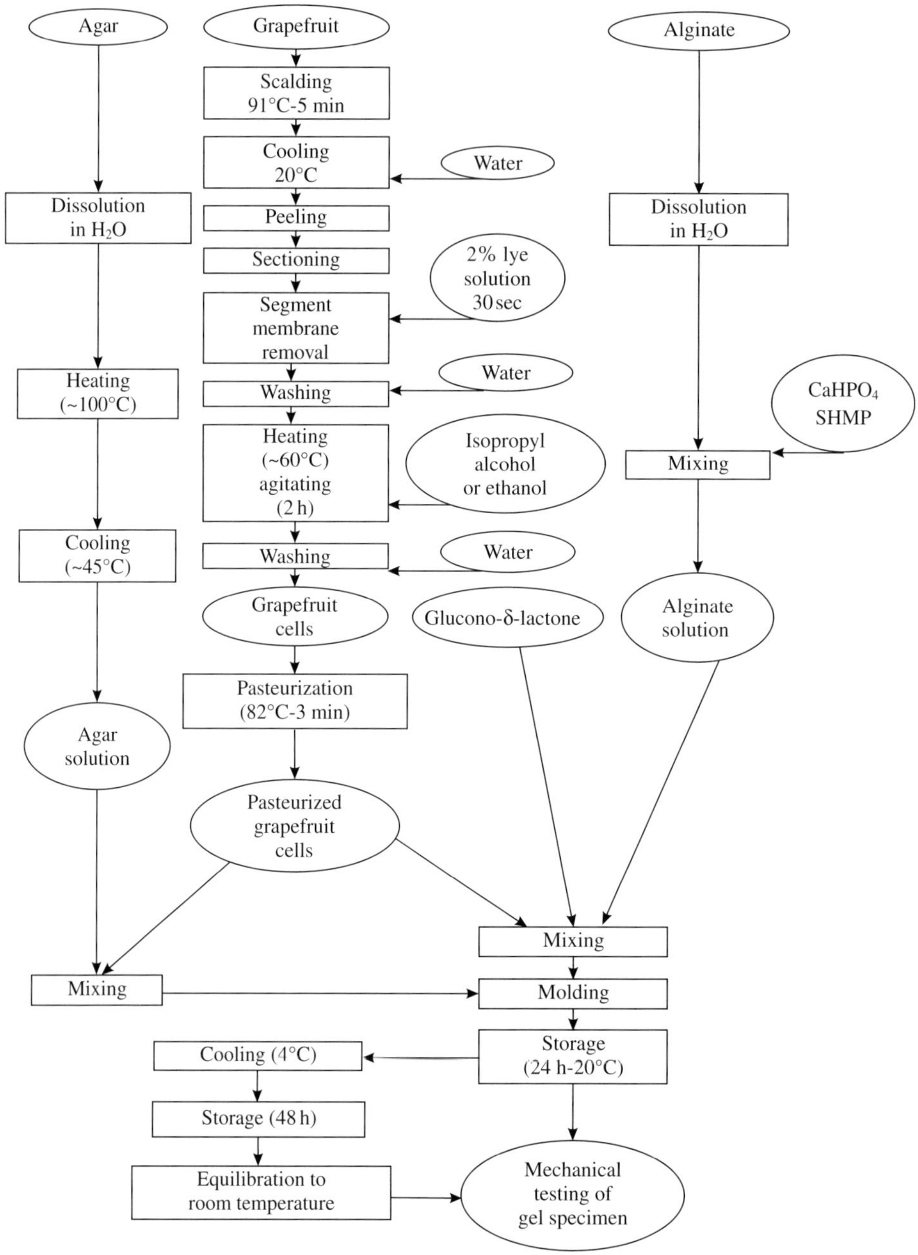

Fig. 7.8 Preparation of succulent texturized grapefruit products. (From Weiner & Nussinovitch, 1994.)

interaction between the juice cells and the surrounding matrix. A significant drop in stress at failure was observed when $50\,g\,kg^{-1}$ cells were introduced into agar or alginate matrices. The strength of an agar matrix with $50\,g\,kg^{-1}$ cells decreased to 43% of the value of the matrix without entrapped cells. For the alginate matrix, the addition of $50\,g\,kg^{-1}$ cells reduced the strength to 34% of its value with no cells present (Weiner & Nussinovitch, 1994).

For both agar and alginate matrices, significant differences in strength occurred upon the addition of up to 150 and $100\,g\,kg^{-1}$ cells, respectively. No significant differences in strength were observed with the addition of higher percentages of juice sacs. A significant drop in deformability modulus already occurred in agar and alginate matrices when $50\,g\,kg^{-1}$ cells were entrapped within the matrix. In other words, the stiffness of the final product decreased with an increase in juice vesicle content. Upon comparing their relative deformabilities, it is clear that agar matrices maintain their stiffness better than alginate matrices. Since the major advantage of using the vesicles is the sensation of sweetened acidic liquid filling the oral cavity when the texturized fruit is consumed, percent weight loss during mechanical testing was determined. In the agar matrix, the addition of juice cells weakened the gel and more liquid was released from the matrix than for a matrix without cells (~7.5–12.5% as compared to ~2% on average) while its mechanical properties were being tested. In the alginate matrix, from a syneresis of 2.2%, values increased to ~3.3–9.4% in the texturized product. Thus percent weight loss in the agar product appeared to be higher than in the alginate product (Weiner & Nussinovitch, 1994).

7.7 Multi-layered texturized fruits

The compressive force–deformation relationships of multi-layered hydrocolloid gels composed of different combinations of agar, four galactomannans, xanthan, carrageenan and konjak mannan, and of gelled texturized fruits based on banana, apple, kiwi and strawberry pulps and agar–LBG combinations, adhered via three different gluing techniques, were calculated from those of the individual layers (Ben-Zion & Nussinovitch, 1997). Fruit pulp (50%) was added to agar–LBG gels to construct gelled, texturized fruits. Apple, strawberry and kiwi pulps had approximately the same influence on the strength of the texturized fruit gels. Stresses at failure were found to be ~28.1–31.8 kPa. The banana pulp considerably decreased stress at failure (14.5 kPa) of the texturized fruit, resulting in the most brittle product. The weakening of gels by the addition of fruit pulp has been extensively studied. In this particular study, banana pulp reduced the stress at failure of 1.5% agar–0.5% LBG gels (~53 kPa), more than the other tested pulps (apple, strawberry and kiwi: 31.8, 28.1 and 29.3 kPa, respectively). At 50% added pulp, there were no significant differences between these three pulps. This reduced influence on gel strength may be dependent on the degree the pulp was milled. From the literature, it is known (Nussinovitch *et al.*, 1991a) that the more thorough the milling, the less disturbance there is to gelled texturized fruit products. The differences between degrees of milling were observed when we pressed some ground (milled) pulp between a white ceramic tile and a transparent glass plate. This simple, convenient and inexpensive method, used in the industry to check the *fineness* of comminuted products, revealed that the banana pulp was ground non-uniformly and contained bigger particles than the other three pulps, hence the resultant disturbance to agar–LBG gels. (For the three failure patterns of such gels, see Chapter 1.) For the four-layered texturized fruits and gels, the compressive deformability of the four layers could be calculated with the same success as that of double layers, with reasonable accuracy over a large range of deformations (Ben-Zion & Nussinovitch, 1996).

Multi-layered hydrocolloid-based foods are important in the framework of foods of the future. The pertinent question here is whether the stiffness of multi-layered texturized fruits can be estimated from the deformability moduli of its individual layers. Recently, a mathematical model to predict the deformability modulus of a four-layered gel array of texturized fruit was developed (Ben-Zion & Nussinovitch, 1996). Each layer was composed of 1.5% agar, 0.5% LBG and one of four pasteurized fruit pulps (50%) – banana, apple, kiwi or strawberry. The model was based on the assumption that the uniaxial stress in the layers is the same and that their deformations are additive. No significant differences were found between experimental and calculated deformability moduli predicted by the model. Thus the deformability modulus of multi-layered texturized fruits can be predicted by using an empirical mathematical model. The model provides a tool for estimating multi-layered gel stiffness and may be applicable to other food systems that behave similarly (Ben-Zion & Nussinovitch, 1996).

7.8 Texturized products for process evaluation

Pilot-plant studies were carried out to evaluate the accuracy and general performance of a new time–temperature integrator (TTI) based on a gelled mixture of alginate–starch and mushroom puree (acting as a model low-acidity food containing large particles) over a temperature range of 115–125°C (Rodrigo *et al.*, 1998). Cylindrical particles of alginate and alginate/starch food puree were developed for use as carriers in a microbiological TTI, to evaluate the high-temperature short-time (HTST) sterilization process. Mechanical properties of the restructured food particle, such as stress and strain at failure, were studied as a function of different composition parameters (alginate and food concentration, pH, type of food added and addition of starch). Addition of food puree (mushroom, meat product, artichoke or potato) produced gels that were weaker than their pure alginate counterparts, although no differences in mechanical properties were obtained among the three levels of food puree concentrations studied – 17, 33 and 50% (w/w). However, the type of food significantly affected these rheological parameters, with artichoke puree producing the weakest gel when added to the alginate. pH also affected mechanical properties – the lower the pH, the weaker the particle. When starch was added, particles that were formed could be frozen without losing their mechanical resistance and handling. Particles containing 2% alginate and 4% starch demonstrated the best mechanical stability (Ocio *et al.*, 1997).

The potential of extrusion to produce value-added restructured peach products was examined. A twin-screw extruder was used to process peach puree (50–100%) with various moisture contents (5–30%), in combinations with starch (gelling agent, used at 0–30%) and water and/or a sugar (0–20%) solution. Barrel temperatures of 75 and 125°C and screw speeds of 100 and 200 rpm were tested. An increase in specific mechanical energy and product temperature increased extrudate hardness. Use of sugar solutions also increased hardness. Moisture content and water activity of extrudates were primarily controlled by the initial moisture content and sugar concentration (McHugh *et al.*, 1996a, b, c).

7.9 Special products related to texturization of vegetative materials

The development of new processed tomato products is seen as a better alternative to reducing tomato production in light of increased tomato yields. Processing techniques and compositional details are reported for products currently in the developmental stage; these are a *hot dog* product made of tomatoes, a mixture of alginate and tomato paste in a wiener-like tube; a tomato chip product comprising a cracker-like texture of tomato paste (~33%), tapioca starch and a little salt, which is fried, and various cake and pie products, e.g. tomato *mince-pie* filling, for the natural health-food market (Hannigan, 1981).

A novel dried texturized fruit-based product (Samarkand mosaic) was developed. It contains a mixture of crushed dried fruits including dried grapes and nuts. The fruit particle size is 2–3 mm (very large in comparison to ground pulp or that achieved from juice filtering). Addition of 2–3% alginate solution or 7–9% gelatin to the fruit, enabled the required thickening and gelling of the texturized products. After mechanical mixing, the mixture is briquetted, dried and packed. It can be used as a dessert and packed in a variety of shapes and forms (Kats & Mikhailenko, 1984).

7.10 Texturized vegetative products based on carrageenan, starch, CMC and konjak mannan

Alginate is perhaps the most commonly used gum for producing texturized vegetative materials (Mitchell & Blanshard, 1976). However, other hydrocolloids have been used for the same purpose, taking into consideration their unique properties. An artificial, edible fruit having a gelled matrix of calcium alginate or calcium low-methoxy pectin (LMP) has dispersed therein gelled particles based on a second gelling agent, e.g. agar, carrageenan or gelatin (Wood, 1975). A novel jelly product with the palatability and flavor of fruit and a process for its manufacture have been described. The jelly, whose main ingredients are fruit juice and konjak flour or konjak mannan, is produced by adding fruit juice to a konjak gel (prepared by heating and cooling an alkaline konjak paste) or to unheated konjak paste at pH 9–10.3, mixing the ingredients and freezing. The product has good keeping quality and taste, aroma and texture when thawed, similar to those of the fruit whose juice it contains (Nozaki & Sakurai, 1992).

Texturized fruits based on peach pulp (10–40%) were produced by adding the pulp to three different polysaccharide gel systems, including 0.75% κ-carrageenan, 0.75% κ-carrageenan plus 0.3% LBG, and 1% alginate. Pulp addition decreased maximum rupture force and deformability modulus values in carrageenan-based gels. No such effects were detected in the alginate-based products (Fiszman & Duran, 1992). Beside the traditional gel-forming agents used for texturization, gelatinous cellulose, prepared by *Acetobacter aceti*, has been reported to have the ability to form fabricated texturized foods (Okiyama *et al.*, 1992). Sweet potato puree was texturized with methylcellulose (MC) or hydroxypropylmethylcellulose (HPMC). The researchers preferred dynamic measurements for the quality assessment of these products. The MC and HPMC samples were significantly tastier than baked roots (Truong & Walter, 1994). Another report investigated texturization of peach puree (0, 15, or 30%) with flour versus starch, creating a product with 65 or 70°Bx via a drying procedure. At similar viscosities, the total solids (TS) level of flour slurries was lower than that of starch slurries. Puree and gelling agents had synergistic effects on gel properties (McHugh *et al.*, 1996a, b, c).

7.11 Unique uses of texturization

7.11.1 Fruit pie fillings

Fillings are spread between layers of cakes or pastry to add flavor and help retain moisture. The filling should not be too moist, to avoid pastry sogginess. The filling should blend with the requested flavor (Gates, 1981). Fruit-pie fillings are very popular. A process for the production of a filling that contains reformed pieces of fruit shaped from fruit puree has been described (Anon, 1978). The process involves putting an alginate *skin* around simulated fruit berries or slices. A fruit filling containing 0.1–2% Ca^{2+} is coated with sodium or potassium alginate while being extruded. The coated pieces then fall into a Ca^{2+} solution and the resulting calcium alginate forms a *skin* over the fruit pieces (Anon, 1978).

7.11.2 Freeze-dehydration and nutritious garnishes

In freeze-dehydration, a frozen material is subject to a pressure below the triple point and heated to cause ice sublimation to vapor. It is a slow and expensive process that is used for the drying of high-value products or to achieve special textures. To use this drying technique to create a simulated fruit gel, a special manufacturing process must be followed (Luh *et al.*, 1976). The fabricated gel system consists primarily of alginate molecules cross-linked with calcium ions. Static compression tests or compression tests using an Instron Universal Testing Machine (UTM) were run to characterize the mechanical properties of cross-linked gels having different compositions. Incorporation of additional components, such as pectin, gelatin or sucrose, to the calcium alginate system modified the textural characteristics of the cross-linked gels. Some hypotheses explaining the effect of added components on gel structure and texture were presented. Comparison of the results obtained from sensory and instrumental methods of texture measurements showed that if the instrumentally measured mechanical properties of the two samples differ by 20% or more, sensory panellists can differentiate between the two samples at a high level of statistical significance (Luh *et al.*, 1977). Freeze-dehydration was also the preferred drying method for the preparation of dried fruit puree as part of a jelly powder formulation. For this product, freeze-dried fruit puree, sugar and thickener (high-methoxylated pectin or modified starch) were used (Ivasyuk & Katun'kina, 1984).

Freeze-dried products are not always preferred, a good example being vegetable garnishes. Nutritious garnishes formulated with textured vegetable proteins are claimed to perform better than dry vegetables in many applications, adding color and crunch at much lower costs than dehydrated or freeze-dried ingredients (Lugay & Kim, 1981). Examples of the variety of such products include: a replacement for expensive, hard to find black Chilean mushrooms, and a look- and taste-alike version of red or green bell pepper dices. Suggested uses for the non-flavored mushroom imitators include dry soups, gravies, sauces, pizza and spaghetti sauces and various processed or dry ethnic-style foods (Anon, 1972).

7.11.3 Texturized fruit and vegetable pieces in breakfast cereals

Extrusion is used to increase the variety of foods in the diet, by producing a range of products with different shapes, textures, colors and flavors from basic ingredients. Several grains have been processed into breakfast cereals. Cornflakes are processed from large maize kernels (grits), with the size of the individual grit determining the size of the final cornflake. Grits are then pressure-cooked for 3 h, dried to 21% moisture, tempered for 2 h to ensure even moisture distribution, flaked, toasted and sprayed with a vitamin solution (Fellows, 1990). Dry fruit pieces are sometimes included in the breakfast cereals. The preparation of formed fruit and vegetable pieces, suitable for inclusion in breakfast cereals, fruit muffins, marshmallow treats, granola bars, ice cream, etc. is described. The fruit or vegetable blend used for the preparation comprises: 7–15% fruit or vegetable solids (dehydrated fruit or vegetables or fresh or frozen concentrates with 40–70% soluble solids); 40–75% sugar; 5–8% water; 3–10% pre-gelatinized starch; 0–30% bulking agent; 0.01–0.05% gum (guar, locust bean gum (LBG) or xanthan), 3–5% oil or fat; and natural colors and flavors. The blend is processed in an extruder at low temperature (60°C) and low shear, at 10 to 100 rpm (Nappen & Koval, 1986). The low-moisture, low-fat, fruit or vegetable morsels have an appearance, texture and taste similar to the fruit or vegetable from which they have been prepared. Low-calorie fruit or vegetable pieces can also be formed by replacing sugar with polydextrose or other suitable artificial sweetener. Examples of fruits and vegetables which can be processed in this way include raspberries, blackberries, apples, pineapples, oranges, lemons, tomatoes, peppers, corn, beets and carrots (Nappen & Koval, 1986). Simulated fruit pieces for inclusion in dry food products can be produced by extrusion of fruit solids, fruit concentrate, thickening agent, edible food-grade acid, sweeteners, coloring, and glycerol. The glycerol/sweetener combination functions as a humectant, which produces water activities of 0.2–0.5 within the product (Lugay *et al.*, 1992). Another report mentioned the fabrication of dried fruit pieces coated with fat to be included in ready-to-eat cereals (Walter & Funk, 1998).

7.11.4 Skin- or sol-coated products

When an alginate solution is dropped into a soluble calcium salt solution, an insoluble calcium alginate *skin* forms almost immediately. This simple reaction provides the basis for the manufacture of many restructured foods. The first patent based on this reaction was granted in 1946 for a process designed to create artificial cherries (U.S. Patent No. 2,403,547), which contained no fruit. Fruit pulp, puree or juice containing soluble calcium salt is extruded to form drops which are coated with a thin skin of alginate or pectate sol. The coated drops are exposed to an aqueous setting bath containing a soluble calcium salt (Sneath, 1975). Artificial fruit products are prepared by mixing a fruit pulp or puree in an aqueous alginate or LMP solution, treating the surface of the mixture with a calcium salt to form a skin and heating the mixture to prevent gelation of the interior (Unilever Ltd., 1976). Another process is described in which an aqueous alginate or pectin sol is co-extruded with a fruit-flavored aqueous composition to form fruit-flavored drops coated with the sol, which are then dropped into an aqueous Ca ion setting bath. Distortion of the drops in the setting bath is minimized by bubbling a gas through the bath to form a foam at its surface (Barwick & Sneath, 1978). A process for the manufacture of shaped foods from fluid materials is based on solidification by heating or homogenization to give a firm or sliceable texture; the product is enclosed in a casing similar to fruit peel, made from a natural material or a nature-identical material, e.g. modified starch (Finnah, 1996).

7.11.5 Texturized potato and onion products for frying and roasting

Foods cook quickly in fat because of the high temperatures involved. Frying is used to alter the eating quality of foods, its main purpose being the development of characteristic colors, flavors and aromas in the crust of the fried foods. This is the reason why so many people enjoy the flavor and texture of these foods. Methods of dry-heat meat cookery include roasting, broiling, pan broiling, pan frying and frying in deep fat (Gates, 1981; Fellows, 1990). Hydrocolloids are used to texturize products for better frying and roasting. A process is described in which a starch-containing dough mixture is extruded in tubular form, encasing a central extrudate of a mixture of onion pieces, modified starch and additives. The extruded compound can be formed into rings or other configurations (Shatila, 1973). Structured potato products incorporating alginate gels have been described. The products may utilize various potato starting materials, e.g. potato flakes, pieces or starch. Duchesse, croquette, French fries and roasted potato products can be produced from the appropriate alginate, with considerable texture control (Anon, 1983). Onion rings are fabricated from minced dehydrated onion by mixing with sodium alginate forming a mash and extruding to make high-walled rings (Katz, 1973). After gelling of the alginate with calcium chloride, the rings are battered and breaded with a white corn-based breading, and deep-fat fried and frozen for packaging in institutional 2-lb packages (Lukas & LaBell, 1983). Another report discusses the manufacturing of extruded gelled foods as a substitute for onion rings. These products have a pre-selected shape and can be produced without breaking apart (Thota & Shah, 1996).

7.12 References

Aguilera, J.M. (1992) Generation of engineered structures in gels. In: (Schwartzberg, H.G. & R.W. Hartel, Eds.), *Physical Chemistry of Foods*. Marcel Dekker Inc., NY, Basel and Hong Kong.

Aguilera, J.M. & D.W. Stanley (1986) In: (Le Maguer, M. & P. Jelen, Eds.), *Food Engineering and Process Applications*. Elsevier Applied Science Publishers, London, **2**, p. 131.

Anon (1972) Textured vegetable garnishes simulate mushrooms, peppers. *Food Processing*, **33**(7), F13.

Anon (1978) Real fruit is used to make simulated fruit. *Food Eng. Int.*, **3**(10), 48–49.

Anon (1980) Alginate fibres create textures of fruits, vegetables and meats. *Food Processing*, **41**(7), 22–24.

Anon (1983) New concept for growing market: structured potatoes. *Food Eng.*, **55**(5), 72–73.

Barwick, B.E. & M.E. Sneath (1978) Process of preparing simulated fruit. United States Patent 4,119,739.

Bellarde, F.B., M.N.H. Jackix & M.A.A.P. da Silva (1995) Development of structural gel of natural passion fruit juice simulating the fruit: sensory evaluation and acceptance. *Ciencia-e-Tecnologia-de-Alimentos*, **15**(3), 225–231.

Ben-Zion, O. & A. Nussinovitch (1996) Predicting the deformability modulus of multi-layered texturized fruit and gels. *Lebensm.-Wiss. u.-Technol.*, **29**, 129–134.

Ben-Zion, O. & A. Nussinovitch (1997) A prediction of the compressive deformabilities of multilayered gels and texturized fruit, glued together by three different adhesion techniques. *Food Hydrocolloids*, **11**(3), 253–260.

Dumas, J. (1984) Process and equipment for producing sticks of fruit jelly. French Patent Application 2,538,681A1.

Fellows, P.J. (1990) *Food Processing Technology*. Ellis Horwood Limited, New York.

Finnah, J. (1996) Process for manufacture of shaped foods. German Federal Republic Patent DE 4411T525 C1.

Fiszman, S.M. & L. Duran (1992) Effects of fruit pulp and sucrose on the compression response of different polysaccharides gel systems. *Carbohydrate Polymers*, **17**, 11–17.

Flora, L.F. & L.R. Beuchat (1979) Dried fruit substitute from Muscadine grape skins. *J. Fruit, Vegetables and Nuts*, **21**(2), 11–13.

Gates, J.C. (1981) *Basic Foods*. Holt, Rinehart and Winston, New York.

Glicksman, M. (1976) Fabricated foods. *Cereal Foods World*, **21**, 17–26.

Hannigan, K.J. (1981) Developing new tomato-paste products. *Food Engineering*, **53**(9), 86–87.

Ivasyuk, N.T. & T.A. Katun'kina (1984) Rapidly soluble jellies obtained by freeze-drying. *Konservnaya-i-Ovoschhesushil'naya-Promyshlennost*, **2**, 24.

Kaletunc, G., A. Nussinovitch & M. Peleg (1990) Alginate texturization of highly acid fruit pulps and juices. *J. Food Sci.*, **55**(6), 1759–1761.

Kats, Z.A. & L.Ya. Mikhailenko (1984) Production technology for Samarkand mosaic. *Konservnaya-i-Ovoshchesushil'naya Promyshelennost*, **11**, 14–15.

Katz, A. (1973) Extrusion process shapes bigger sales for frozen onion rings. *Quick Frozen Foods*, **35**(7), 32–34.

Less, R. (1998) Ingredients used in the manufacture of sugar confectionery jellies. III. Jellies from marine sources. *Confectionery Production*, **64**(5), 14–15.

Lillford, P.J. (1985) In: (Simatos, D. & J.L. Multon, Eds.), *Properties of Water in Foods*. Nijhoff Publisher, Dordrecht, p. 543.

Lodge, N. (1981) Kiwi fruit: two novel processed products. *Food Technol. New Zealand*, **16**(7), 35, 37–38, 41.

Lugay, J.C. & M.K. Kim (1981) In: (Stanley, D.W., E.D. Murray & D.H. Lees, Eds.), *Utilization of Protein Resources*. Food and Nutrition Press, Westport, CT, p. 177.

Lugay, J.C., J.L. Newkirk, K. Morimoto & P.K. Roy (1992) Process for making simulated fruit pieces. United States Patent 5,084,296.

Luh, N., M. Karel & M.J. Flink (1976) A simulated fruit gel suitable for freeze dehydration. *J. Food Sci.*, **41**, 89–93.

Luh, N., J.M. Flink & M. Karel (1977) Fabrication, charcterization, and modification of the texture of calcium alginate gels. *J. Food Sci.*, **42**(4), 976–981.

Lukas, S. & F. LaBell (1983) Fabricated onion rings have homestyle look and mild onion flavor. *Food Processing*, **44**(9), 26–27.

Matz, S.A. (1984) *Snack Food Technology*. AVI Publishing Co., Wesport, CT.

McHugh, T.H., J. Hsu, C.C. Huxsoll & G.H. Robertson (1996a) Effects of peach puree concentration, starch concentration and total solids on the properties of restructured peach products. In: *IFT Annual Meeting: Book of Abstracts*, p.114.

McHugh, T.H., J. Hsu, C.C. Huxsoll & G.H. Robertson (1996b) Extrusion processing of peach puree-based restructured fruit products. In: *IFT Annual Meeting: Book of Abstracts*, p.183.

McHugh, T.H., J. Hsu, C.C. Huxsol & G.H. Robertson (1996c) Effects of flour versus starch on the properties of restructured peach puree based gels. In: *IFT Annual Meeting: Book of Abstracts*, p. 89.

Mitchell, J.R. & J.M.V. Blanshard (1976) Rheological properties of alginate gels. *J. Texture Studies*, **7**, 219–234.

Mouquet, C. (1996) Assessing fruit quality through texture measurements. *Fruits*, **5**, 307–315 (in French).

Mouquet, C., J.C. Dumas & S. Guilbert (1992) Texturization of sweetened mango pulp: optimization using response surface methodology. *J. Food Sci.*, **57**(6), 1395–1400.

Mouquet, C., C. Aymard, S. Guilbert, G. Cuvelier & B. Launay (1997) Influence of initial pH on gelation kinetics of texturized fruit pulp. *Lebensm.-Wiss. u.-Technol.*, **30**, 129–134.

Nappen, B.H. & P.L. Koval (1986) Blends suitable for the preparation of formed fruit or vegetable pieces for food products. European Patent Application EP 0204939 A2.

Nozaki, H. & S. Sakurai (1992) Jelly resembling the flesh of fruit. United States Patent 5,089,285.

Nussinovitch, A. (1993) Gum-based texturized products. In: *Yearbook of Science and Technology*, McGraw-Hill, NY, pp. 138–140.

Nussinovitch, A. (1997) In: *Hydrocolloid Applications, Gum Technology in the Food and Other Industries*, Blackie Academic & Professional, London, UK, pp. 328–338.

Nussinovitch, A. & M. Peleg (1990) Mechanical properties of a raspberry product texturized with alginate. *J. Food Process. Preserv.* **14**, 267–278.

Nussinovitch, A., M. Peleg & M.D. Normand (1989) A modified Maxwell and a nonexponential model for characterization of the stress relaxation of agar and alginate. *J. Food Sci.*, **54**, 1013–1016.

Nussinovitch, A., I.J. Kopelman & S. Mizrahi (1991a) Mechanical properties of composite fruit products based on hydrocolloid gel, fruit pulp and sugar. *Lebensm.-Wiss. u.-Technol.*, **24**, 214–217.

Nussinovitch, A., I.J. Kopelman & S. Mizrahi (1991b) Modeling of the combined effect of fruit pulp, sugar and gum on some mechanical parameters of agar and alginate gels. *Lebensm.-Wiss. u.-Technol.*, **24**, 513–517.

Ocio, M.J., S.M. Fiszman, F. Gasque, M. Rodrigo & A. Martinez (1997) Development of a restructured alginate food particle suitable for high temperature short time process validation. *Food Hydrocolloids*, **11**(4), 423–427.

Okiyama, A., M. Motoki & S. Yamanaka (1992) Bacterial cellulose. II. Processing of the gelatinous cellulose for food materials. *Food Hydrocolloids*, **6**(5), 479–487.

Pelaez, C. & M. Karel (1981) Improved method for preparation of fruit-simulating alginate gels. *J. Food Process. Preserv.*, **5**, 63–81.

Ring, S.G. & G. Stainsby (1982) In: (Phillips, G.O., D.J. Wedlock & P.A. Williams, Eds.), *Progress in Food and Nutrition Science*, vol. 6, *Gums and Stabilisers for the Food Industry: Interactions of Hydrocolloids*. Pergamon Press, Oxford, pp. 323–329.

Rodrigo, F., M.C. Rodrigo & A. Martinez (1998) Evaluation of a new time temperature integrator in pilot plant conditions. *Food Res. Technol.*, **206**(3), 184–188.

Shatila, M.A. (1973) Fabricated onion ring, United States Patent 3,761,282.

Silberberg, A. (1989) In: (Glass, J.E., Ed.), *Polymers in Aqueous Media: Performance Through Association. Advances in Chemistry Series No. 223*, American Chemical Society, Washington, DC, p. 3.

Sneath, M.E. (1975) Simulated soft fruits, United States Patent 3,922,360.

Szczesniak, A. (1968) Simulated fruits and vegetables, United States Patent 3,362,831.

Tateo, F. (1985) Procedures for increasing utilization of fruit products. *Industrie Alimentari*, **24**(228), 529–532.

Thota, H.A.P. & C.S. Shah (1996) Process for extruding gelled product. United States Patent 5,578,337.

Tolstoguzov, V.B. (1971) Method of preparing artificial foodstuffs. USSR Patent 296,554.

Tolstoguzov, V.B. & E.E. Braudo (1983) Fabricated foodstuffs as multicomponent gels. *J. Texture Studies*, **14**, 183–212.

Truong, V.D. & W.M. Walter, Jr. (1994) Physical and sensory properties of sweet potato puree texturized with cellulose derivatives. *J. Food Sci.*, **59**(6), 1175–1180.

Truong, V.D., W.M. Walter, Jr. & F.G. Giesbrecht (1995) Texturization of sweet potato puree with alginate: effects of tetrasodium pyrophosphate and calcium sulfate. *J. Food Sci.*, **60**(5), 1054–1059, 1074.

Uniliver Ltd. (1974) Simulated fruit. British Patent 1,369,198.

Uniliver Ltd. (1976) Artificial fruit product. British Patent 1,428,362.

Walter, D.L. & D.F. Funk (1998) Fabricated fruit pieces and method of preparation. United States Patent 5,718,931.

Weiner, G. & A. Nussinovitch (1994) Succulent, hydrocolloid-based, texturized grapefruit products. *Lebensm.-Wisse. u.-Technol.*, **27**(4), 394–399.

Wood, F.W. (1975) Artificial fruit and process therefore. United States Patent 3,892,870.

Chapter 8

Hydrocolloid Cellular Solids

8.1 Introduction

The word cellular solid originates from the word *cell*, which in turn is derived from the Latin *cella* – a small compartment, an enclosed space. Connecting the solid edges is a simple way to fill space with unique structures, which can be observed in many natural instances, e.g. wood, cork, sponge and coral (Gibson & Ashby, 1988). Man-made natural-based or synthetic cellular products include items from disposable coffee cups to crash padding in aircraft cockpits. Foamed polymers, metals, ceramics and glass have been used for insulation, cushioning and absorption of impact kinetic energy (Nussinovitch, 1997). Cellular solids contain different ordered and disordered structures. Of major importance is the distinction between open-cell (inter-connected) and closed-cell (a cell sealed off from its neighbors by membrane-like faces) cellular solids (Gibson & Ashby, 1988; Jeronomidis, 1988).

8.2 General applications of cellular solids

The most important properties of cellular solids are density, conductivity, Young's modulus and strength. A wide range of these properties can be achieved with different structures, leading to many possible applications. Low densities mean light, stiff, large portable structures with flotation ability. Low thermal conductivity leads to affordable thermal insulation. This property can be used in cars, ships, buildings and trucks. Glass foams can be used in cases of possible fire hazard or for extending the lifetimes of pipes and roved products. Packaging with cellular solids is used to absorb impact energy as a result of its ability to pass through large strains without generating high stresses. Low density leads to a light package, which in turn reduces shipping costs. Natural-based (like balsa) or synthetic (cellulose acetate foam) cellular solids can be used in sandwich panels (Gibson & Ashby, 1988). Another use of closed-cell plastic cellular solids is their potential buoyancy. The buoyancy factor (B) is calculated by the simple equation

$$B = \frac{(\rho_{\text{water}} - \rho_{\text{foam}})}{\rho_{\text{water}}} \tag{1}$$

Other uses might be for filtration, where metal is poured through open-cell ceramic foam to achieve high-quality metal castings. Foam sheets can be used as carriers for inks, dyes, lubricants and enzymes for chemical processing (Gibson & Ashby, 1988).

8.3 Structure of cellular solids

The single most important structural characteristic of a cellular solid is its relative density (ρ^*/ρ_s): the density, ρ^*, of the foam divided by that of the solid from which it is made, ρ_s. The fraction of pore space in the foam is its porosity, calculated by $[1 - (\rho^*/\rho_s)]$. Cellular solids have relative densities less than 0.3; many can have much lower values. Cell size is less important since most mechanical and thermal properties depend only weakly on cell size (Gibson & Ashby, 1988). Equi-axed cells confer isotropic properties on the cellular solid. Flattened or slightly elongated cells confer properties that are dependent on direction. Cells can be two-dimensional (cell walls have a common generator), or three-dimensional (randomly oriented in space), within which there is distinction between open and closed cells in the solid structure. Cell wall connectivity (number of faces that meet an edge) can be from 3 up to 6 (Gibson & Ashby, 1988). Ordered structures (as in the case of bee honeycombs) and disordered three-dimensional networks can be found side by side in the world of cellular solids. Aside from open versus closed cells, the structure can be classified according to its flexible versus rigid (or brittle) cell walls and cell wall distribution, thickness, shape and uniformity (Gibson & Ashby, 1988; Peleg, 1997).

8.4 Edible cellular solids

Many foods are foams. Bread usually consists of closed cells, expanded by the fermentation of yeasts or by CO_2 from bicarbonate. Breads display an enormous range of phase volumes, air-cell dimensions and anisotropy. Meringue consists of foamed egg white and sugar. The size of the gas cells, the maximal included phase and ease of processing are influenced by the content and character of the surface-active agents used (Nussinovitch, 1997). Foamed chocolate represents a food which has been expanded to change its texture. Lillford (1989) described an X-ray projection of aerated chocolate. The air-phase volume included within the bar is composed of interconnecting air cells of various sizes. This has a major influence on the bar's mechanical properties. In products such as sugar confections, the air-phase volume is high, and cell walls are thin and composed of very brittle sugar glass. The air is the primary cause of the textural difference between boiled sweets and this foamed structure, although both are made from sugar and water (Lillford, 1989). In fact, almost any material capable of solidification can be aerated. Other hard brittle candies are also often expanded to make them attractive to consumers or, if they are sold by volume, to make them less expensive. Other important foods are breakfast cereals and snack foods, which are steam-foamed to produce a different texture and crunchiness.

There are two major types of foams. The first, in terms of mechanical properties, is defined when the geometry of the structure, due to the inclusion of a second phase, does not influence the basic deformation mechanism of the matrix material (Nussinovitch, 1997). Examples are high-density foams with small included phase volumes, or structures with lower densities built of very brittle material, where the deformation prior to fracture is very small (Lillford, 1989). Another report describes sponge preparation by freezing and thawing amylose gels. The sponges have a characteristic *leafy* structure reminiscent of the fleshy part of some seafoods. The mechanical properties of the sponges were evaluated by uniaxial relationship of the sponge and could be described by $\sigma = E_1\varepsilon + E_2\varepsilon^n$, where σ and ε are the true stresses and strains, E_1, E_2 and n are constants. It has been suggested that the first component is the contribution of structural compaction which becomes dominant at large deformation levels. The effects of consecutive compressive cycles as well as amylose concentration and freezing methods were also evaluated in terms of mechanical parameters. An explanation of the latter has been suggested in terms of the deformation mechanism of sponges (Torres *et al.*, 1978). Another edible sponge previously described consisted of 45% (by weight) maltodextrin with a mean polymerization degree of 15–30 glucose units, 45% eggs and 10% water (Nejedly *et al.*, 1986).

Most food foams have larger phase volumes and make up the second foam type: those in which the geometry influences the deformation mode and therefore influences the modulus and strength of the structure. The properties of sponge cakes were studied by Attenburrow *et al.* (1989). They have an open-cell structure. Water is included within the cell walls, which by their presence and content influence the mechanical properties. In other words, the mechanical properties can only be determined if the products being tested have been equilibrated in an atmosphere of controlled relative humidity. For sponge cakes, major changes in the stress–strain curve are observed if samples have been equilibrated at $Aw < 0.33$ and $Aw > 0.57$ (Attenburrow *et al.*, 1989). At lower moisture contents, non-recoverable brittle failure occurs, whereas at higher moisture contents, the mode of collapse appears to be recoverable elastic buckling. Samples showed a square dependence of both modulus and fracture stress on bulk density. The plasticizing effect of the moisture could be concluded from the proportionality constants for both parameters. Among the parameters studied, the modulus of the samples showed the highest statistically significant correlation to sensorially perceived hardness (Nussinovitch, 1997).

Some foods are composed mainly of protein, starch and sugar. These ingredients have solid densities ranging from 1.5 to 1.6. In most cases, the apparent density can be used for edible cellular solids even without the common normalization procedure, when foams of the same composition are being compared. Less straightforward is the quantification of mechanical properties, such as stiffness (by Young's modulus). Therefore, solid foods cannot be approximated as elastic moieties, even as a first guess (Peleg, 1997).

8.5 Compression of cellular solids

Cellular solids compressed to small deformations return to their original shape after load removal, demonstrating their elastic properties. Deformability modulus can be calculated from the linear or approximately linear engineering stress–strain relationships (Attenburrow *et al.*, 1989; Meincecke & Clark, 1973; Phillips & Waterman, 1974; Warburton *et al.*, 1990). Theoretical analysis of ideal solid foams showed that the modulus and density have a power law relationship (Gibson & Ashby, 1988). For open, inter-connected cells

$$E/E_s = k\,(\rho^*/\rho_s)^2 \tag{2}$$

and for closed cells:

$$E/E_s = k'(\rho^*/\rho_s)^3 \tag{3}$$

where E and E_s are the moduli of the foam and cell wall material respectively, ρ^* and ρ_s are their corresponding densities, and k and k' are the proportionality constants. Agreement between the theoretical relationships and the actual behavior of sponge cakes and starch foams was reported by Attenburrow *et al.* (1989) and Warburton *et al.* (1990).

The magnitude of the power cannot be accurately determined, due to experimental scatter. Calculations give values that are relatively close to those derived from theory. In the case of large deformations of cellular spongy materials (with open or closed cells), compressibility is determined by three mechanisms. At small deformation (see above), the cell's geometry changes slightly and the structure is more or less elastic (Peleg, 1997). At higher deformations, cell collapse occurs (as a result of fracturing or buckling), and closed cells can be punctured or burst. At this stage, the force remains more or less steady due to little or no additional resistance. At a certain deformation level, the remaining space is filled with collapsed and/or fractured cell wall material and the mechanical resistance of the collapsed structure increases dramatically. At this stage the compressed material resembles an incompressible solid (Peleg, 1997).

8.6 Models for describing stress–strain behavior

Cellular solids are highly compressible. Their cross-sectional area remains almost unchanged even after being compressed to large deformation. In other words, engineering stress–strain relationship can legitimately be used. The engineering stress σ_E, and strain ε_E are defined as:

$$\sigma_E = F/A_0 \tag{4}$$

$$\varepsilon_E = \Delta H/H_0 \tag{5}$$

where F is the force, A_0 and H_0 the specimen's initial area and height respectively, and ΔH is the absolute deformation. Figure 8.1 shows a typical compressive stress–strain curve for a cellular material, composed of three regions. The first is a linear-elastic region, followed by a plateau of roughly constant stress (second region), leading to a final region of steeply rising stress. Each region is associated with a particular

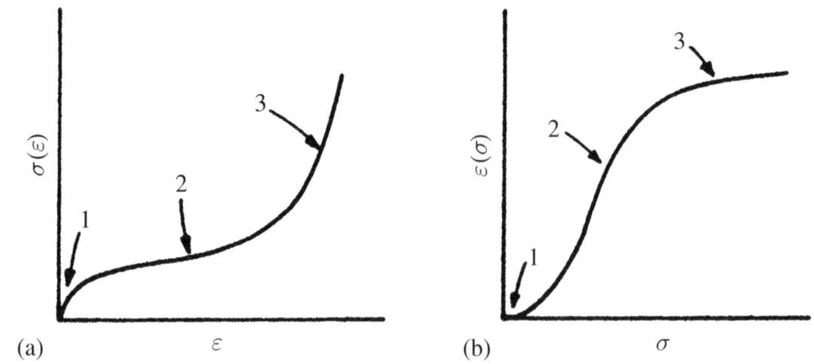

Fig. 8.1 Schematic representation of the typical compressive stress–strain (a) and strain–stress (b) relationships of sponges. 1, deformation of the original matrix; 2, densification; 3, compaction of the collapsed cell wall material.

deformation mechanism. When a specimen is loaded the cell walls are bent, giving linear elasticity if the cell wall material is linear-elastic. When a critical stress is reached, cells begin to collapse and eventually, at high strains, collapse is sufficient to allow opposing cell walls to touch (or their broken fragments to pack together), and further deformation compresses the cell wall material itself. This gives the final, steeply rising portion of the stress–strain curve and is called densification (Nussinovitch, 1997). The typical sigmoidal shape of the curve is maintained if the strain is presented as Hencky's (natural) strain:

$$\varepsilon_{\mathrm{H}} = -\ln[H_0/(H_0 - \Delta H)] \tag{6}$$

which is an indication of the true compressibility (in incompressible material, the upward concavity of the curve disappears when the cross-sectional area's expansion and the non-linearity of the strain are accounted for) (Peleg, 1997).

The compressive stress–strain curve, up to ~80% deformation, can be described by a variety of empirical mathematical models, among them (Peleg *et al.*, 1989; Swyngedau *et al.*, 1991a, b):

$$\sigma(\varepsilon) = C_1\varepsilon/[(1 + C_2\varepsilon)(C_3 - \varepsilon)] \tag{7}$$

and

$$\sigma(\varepsilon) = C_1'/[\varepsilon/(C_3 - \varepsilon)]^n \tag{8}$$

$$\sigma(\varepsilon) = -(1/C_1'')\ln[1 - (\varepsilon/(C_3'')^{n''})] \tag{9}$$

and

$$\sigma(\varepsilon) = C_1'''\varepsilon^{n_1'} + C_2'''\varepsilon^{n_2'} \quad (n_1 < 1,\ n_2 > 1) \tag{10}$$

where the Cs and ns are constants. The fit of these empirical models to the experimental data, gathered when two edible cellular solids (breads) were compressed was, in fact, excellent (Swyngedau *et al.*, 1991a). The calculated constants can serve for quantitative comparison between the curves of different solid foams and of the same material when it is subjected to repeated compression–decompression cycles (Peleg *et al.*, 1989), to calculate the compressibility pattern of layered arrays of sponges (Peleg, 1993; Swyngedau *et al.*, 1991b). Due to changes in the constants during processing, they can serve as a sensitive tool to study the effects of drying, storage, heating and humidification on the texture of the spongy product.

8.7 Elastic properties of edible cellular materials

Elasticity is the ability of a body to return to its original shape after removal of the deforming load (Peleg, 1997). Thus, 100% recovery is considered to be 100% elasticity, and 0% recovery is considered to be 100% plasticity. An accurate way of determining the degree of elasticity is to calculate the ratio between the areas under the decompression and compression curves (Kaletunc *et al.*, 1991; Nussinovitch *et al.*, 1992a). When a cellular spongy material is compressed, irreversible structural changes can occur as a result of cell wall rupturing and/or opening of closed cells, the extent of which depends on the imposed strain. Degree of elasticity (Fig. 8.2) can be dependent on strain and is also a function of the number of compression–decompression cycles imposed on the specimen (Kaletunc *et al.*, 1992). In bakery products, the sigmoidal shape of the stress–strain curve disappears after the first compressive cycle. With regard to the previously presented models, the loss of the *shoulder* and transformation of the shape into one with ever-increasing upward concavity is expressed by $C_2 = 0$ in Equation 7, and $n_1 > 1$ in Equation 10. Similar behavior has been reported for synthetic foams, although a vestige of the sigmoidal shape is still detectable in the decompression curve (Peleg *et al.*, 1989).

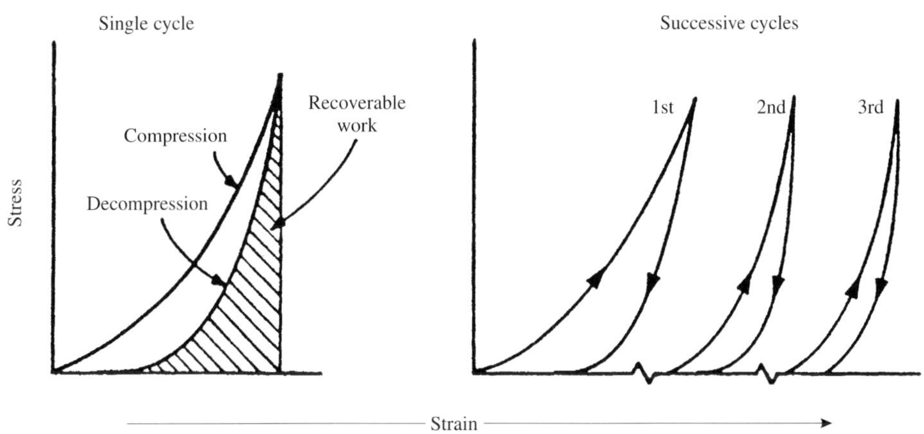

Fig. 8.2 Schematic view of stress–strain relationships of marshmallows subjected to repeated compression–decompression cycles. The changes in the shape of the curve are probably due to rupture of closed cells. (From Kaletunc *et al.*, 1992, with permission from Food & Nutrition Press, Inc.)

8.8 Layered sponges and compressibility of spongy particulates

Different cellular materials can be arranged in layers of the same or different thicknesses. Such arrays can be compressed uniaxially and their cross-sectional area is assumed to be unchanged. The stress along the array can be considered to be the same in all the layers whereas the total deformation may be represented as the sum of the deformations of each layer (Peleg, 1997). Mathematically this is expressed as:

$$\sigma_{\text{total}} = \sigma_i \tag{11}$$

where σ_{total} is the array's stress and σ_i is the stress in an individual layer i, and

$$\varepsilon_{\text{total}} = (1/H_{\text{total}}) \sum H_{0i}\varepsilon_i(\sigma) \tag{12}$$

where $\varepsilon_{\text{total}}$ is the array's strain, ε_i is the individual layer's thickness, and $\varepsilon_i(\sigma)$ is its strain as a function of the stress. The array's initial overall thickness, $H_{0\text{total}}$ is the sum of that of the individual layers, i.e.:

$$H_{0\text{total}} = \sum H_{0i} \tag{13}$$

Equations 6 and 7 are especially convenient as deformability models for sponges in that they can be used to express the strain as an explicit algebraic function of the stress, $\varepsilon(\sigma)$. Inserting the terms $\varepsilon_i(\sigma)$ and the corresponding H_{0i} into Equation 12 enables a calculation of the stress–strain relationship of any layered array of sponges, as long as the assumption that the cross-sectional area remains practically unchanged remains valid (Swyngedau & Peleg, 1992; Swyngedau *et al.*, 1991a, b). Information on how to avoid problems related to requiring close guesses in the non-linear regression procedure, and how to use a polynomial model to eliminate the difficulty, can be found elsewhere (Peleg, 1993).

The assumption that the cross-sectional area of the compressed object is uniform and remains practically unchanged cannot be made for spherical spongy particulates or those having an irregular shape, such as puffed popcorn (natural or synthetic), or other puffed or dehydrated food particulates (Nuebel & Peleg, 1993, 1994), even though the same three deformation mechanisms control the compressibility patterns. In addition, the specific geometry is a major factor. Thus the force–deformation curves of individual particles differ from one another and can also lose their sigmoidal shape (Nussinovitch *et al.*, 1991), as happens with bulk compressibility.

8.9 Brittle foams

Dry cereals and snacks are edible foams characterized by their brittleness. Their force–deformation curve is irregular and jagged, suggesting small and large fluctuations in force. Sometimes a general sigmoidal shape is still discernible (e.g. Attenburrow *et al.*, 1989; Barrett & Peleg, 1992; Nussinovitch, 1995; Nussinovitch *et al.*, 1993).

In many cases, the curves are too jagged to reveal any characteristic general shape. This is expected with particulates having non-uniform structure or irregular geometry (Harris & Peleg, 1996; Wollny & Peleg, 1994). Cell wall materials and cell size distribution are responsible for these products' mechanical properties (Barrett & Peleg, 1992; Warburton *et al.*, 1990). The moisture within the product has a major influence on its brittleness and deformability pattern (Peleg, 1997). Sorption of moisture invariably reduces the brittleness of the structure (Barrett *et al.*, 1992; Harris & Peleg, 1996; Wollny & Peleg, 1994), and the jagged force–deformation curve becomes smoother. The brittleness of cellular snacks and puffed cereals can be quantified in terms of the degree of jaggedness of their force–deformation curves. The degree of jaggedness itself can be expressed in terms of statistical measures, the mean magnitude of the Fourier transform power spectrum, and/or the apparent fractal dimension of the normalized original curve (Peleg, 1997). It is interesting to note that the force–deformation curve of a brittle particulate is much smoother than that of an individual particle, as a result of an averaging effect and cushioning.

8.10 Hydrocolloid cellular sponges

8.10.1 Sponges made by drying gels with and without internally produced gas bubbles

Mechanically stable solid sponges were produced by freeze-drying 2% agar or 1% alginate gels. The sponges had characteristic compressive stress–strain curves that could be described by a three-parameter model originally developed for polymeric sponges and bakery products. Internally produced bubbles, formed by immersing bicarbonate-containing gels in an acid bath, resulted in a considerable loss of mechanical integrity in the dry agar sponges but not in those from alginate (Nussinovitch *et al.*, 1993).

Alginate can be used to create gels filled with internally produced CO_2 gas bubbles: cold-set alginate gels including $CaCO_3$ are produced by immersing the gels in citric acid (various concentrations) solutions, the volume of which should be sufficient to guarantee excess acid, which then diffuses into the gels. Bubbles are formed and their number can be counted using a light microscope. Gas bubbles seem to be trapped within the gel body and bulge from its outer surface. The motion of the acid within the gel is controlled diffusion, as evidenced by the linearity of the penetrated distance versus time$^{1/2}$. After 2.5 h, about 900 bubbles cm^{-3} were counted. This number increased to about $2.5–2.7 \times 10^3$ after 24 or 36 h, depending on the carbonate concentration. Bubble formation decreased the density of the gels, causing them to float. After a while, bubbles began leaving the gel, causing some damage to its integrity, and allowing the liquid to gradually fill the empty spaces. Consequently, the gels began to sink again. Alginate gels without carbonate immersed for 2.5 h in a 0.5% citric acid solution increased their average stress at failure from 28 to 46 kPa. The Hencky's strain at failure of these gels increased from 0.64 to 0.83, indicating that the gels become less brittle. The increase in failure stress and strain was found for all tested alginate systems immersed in 0.5–2.0% citric acid solutions. This may have

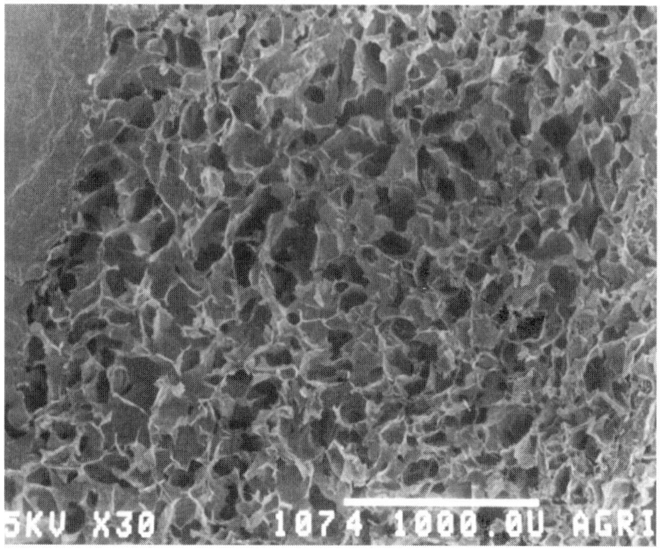

Fig. 8.3 Typical appearance of a freeze-dried gel specimen. The shape of the dried specimen approximates that of the gels before dehydration, thus products can be designed in any shape or size, by building the desired molds. (From Nussinovitch, 1995.)

been due to acid-induced cross-linking which helped the gel retain its mechanical strength, even in the face of the structural disruption caused by bubble formation. The presence of carbonate, however, had a disruptive effect, primarily manifested in lower stiffness. Thus gel strength depended on both acid and $CaCO_3$ concentrations (Nussinovitch *et al.*, 1992a, b).

After drying, dry cellular solids are produced (Fig. 8.3). Because the shape of the final product approximates that of the gels before dehydration, products can be designed in any shape or size, by building the desired molds. The structure of the cellular solid is determined by the drying process as much as by changes in the gel's processing and components. In alginate cellular solids, immersing the gels in an acid bath did not result in a drastic loss of mechanical integrity. This appears to be because the disruptive effects of bubble formation were at least somewhat offset by the more extensive cross-linking (Nussinovitch *et al.*, 1993, 1995).

8.10.2 Sponges produced by fermentation of immobilized yeasts

Yeasts were immobilized within an agar gel: the higher the concentration of the microorganisms, the higher the disturbance to the gel's integrity. In other words, stresses at failure and deformability moduli decreased after immobilization of the yeasts. The brittleness of the gels also increased (their strain at failure decreased). Similar results have been achieved for bacterium, yeast and spore immobilization (Nussinovitch *et al.*, 1994). After yeast immobilization, the gels were immersed in an excess volume of 5% sucrose solution. Slow fermentation occurred, perhaps because no nitrogen source was added. A longer lag time for yeast growth may also have caused immobilized yeast to react later than their non-entrapped counterparts.

Fig. 8.4 SEM micrograph of a yeast sponge showing the yeasts on and entrapped within the sponge walls.

As a result of the fermentation, CO_2 bubbles and ethanol were produced and the pH decreased. The gas bubbles moved from their site of production (inside the gel) to the surface of the gel, causing some mini-cracks and subsequently influencing the structure of the resultant sponge. The longer the fermentation, the less strong and stiff the gels. The gels were freeze-dried to produce cellular solids (Fig. 8.4). All the sponges showed a sigmoidal stress–strain curve characteristic of cellular solids, which is a manifestation of the three aforementioned deformation mechanisms. By applying Equation 8 to the stress–strain curves and calculating their constants by non-linear regression, a major difference between sponges was observed in the magnitude of C_1. The higher the concentration of the entrapped microorganisms, the smaller the value of this constant, which serves as a scale factor for stress. Yeast cells were distributed in and on the cell walls of the sponge and were attached to its outer surface. The compression of all sponges produced after 3 days of immersion in sucrose solution resulted in stress–strain relationships similar to those of a regular sponge. This was true for all yeast concentrations used in the study (Nussinovitch & Gershon, 1997b). After 7 days immersion in the sucrose solution, a different phenomenon was observed. For a 10^7-yeasts per g gel, a *regular* stress–strain relationship was still observed, whereas higher initial yeast concentrations, such as 10^8 and 10^9 yeasts per g gel, produced materials which did not resemble sponges in their stress–strain behavior. This can be partially explained by noting that an increase in the time of fermentation results in an increase in biomass (a fivefold increase in the protein content of the gels before freeze-dehydration was observed). In compressed sponges with initially high yeast concentrations, after 7 days immersion in sucrose solution, compaction of a *yeast-hydrocolloid* rather than a hydrocolloid network occurred, resulting in different products and properties. Comparing sponges prepared from gels without yeast to those prepared from gels with 10^9 entrapped yeast per g showed a decrease in porosity from ~96% to ~92%. This may be due to the increase in sponge dry matter content.

8.10.3 Oil gels and sponges

Oil was included in the sponges to change properties such as structure, density and porosity. First, oil was included within the alginate gels. The higher the content of the oil within the gel, the lower its stress at failure and stiffness, as reflected by the deformability modulus, and the smaller the Hencky's strain at failure; in other words, the gel was more brittle. Two systems of oil gels and sponges were studied: in the first, gels with and without oil were simply freeze-dried directly; in the second, the gels were heat-treated at 85°C for 15 min in water, three times in succession. Each time the water and the extracted oil were discarded. Oil in the gels and sponges was estimated by the Soxhlet method. After heat treatment, 40–50% of the oil had *left* the gel. After freeze-dehydration, the oil percentage within the sponge increased. Oil sponges showed ruggedness in their stress–strain curves. The higher the oil content within the sponge, the smoother the curve; in addition, C_3 from Equation 8 decreased in parallel to the increasing oil content. The higher the bulk density of the sponge, the more the stress tended to steepness at smaller deformations. After extraction, the stress–strain curves became more ragged. The heat treatment may have disrupted the gel structure, physically damaging the specimen surface. The porosity of the sponges changed dramatically after oil inclusion: porosities of the resultant cellular solids decreased from ~95% (no oil included) to ~80% (40% oil included in the gel before dehydration). In addition, the higher the oil content in the sponge, the more closed cells there were within its structure (Fig. 8.5). Moreover the structure of the cells changed from big openings to rounder, smaller ones. The oil could be detected as mini-drops embedded within the solid wall of the matrix (Nussinovitch & Gershon, 1997a).

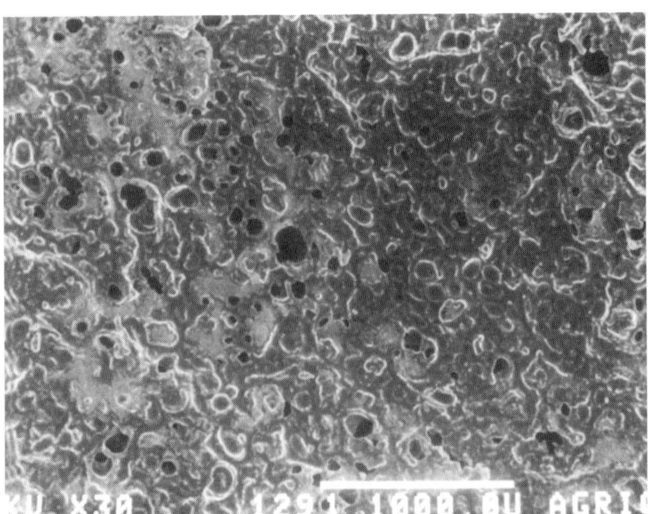

Fig. 8.5 Hydrocolloid sponge with olive oil included within its matrix. The higher the oil content the more changes within the cellular solid structure and the more closed cells there are.

8.10.4 Enzymatically produced hydrocolloid sponges

Sponges with a cellular structure were created by subjecting agar–starch gels to α-amylase activity prior to freeze-dehydration. Various starch concentrations (0.5–1.5%), enzyme concentrations (1000–1500 ppm) and times of exposure to the enzyme were selected to change the structure and mechanical properties of the hydrocolloid cellular solid. The influence of the duration of enzymatic treatment on the cellular structure of freeze-dried 2% agar–1.5% starch gel specimens (sponges) was studied. The enzyme (1500 ppm of α-amylase) solution was incubated with the agar–1.5% starch gels at 55°C for 24 and 72 h. The enzyme diffused into the gels and began to decompose the substrate. The hydrolysis of the starch caused enlargement of the pores within the gel matrix and consequently within the sponge. As a consequence of this process, the dry sponges underwent major structural changes: the longer the exposure to the enzyme, the larger their pores. These changes may have contributed to changes in the mechanical properties of the sponges. The effect of increasing enzyme concentrations was only slightly detectable: for the highest enzyme concentration (1500 ppm) the stress–strain curve was located only slightly lower than those of the control and 1000 ppm enzyme concentration (Nussinovitch *et al.*, 1998).

The influence of 1500 ppm α-amylase on agar gels containing 1.5% starch, at incubations of 24 or 72 h, was checked. The longer the immersion time, the greater the possibility of starch decomposition, resulting in a weaker sponge. Furthermore, the sponges became slightly more porous after 24 and 72 h (~0.93–0.96) than the control sponges (~0.87–0.91). It should be noted that incubation at 55°C for 72 h, even without enzyme activity, led to a decrease in the sponge's mechanical properties, as could be observed by comparing the stresses measured at a strain of ~0.2, at the shoulder. Starch degradation was verified by HPLC analysis. For agar–1.5% starch gels incubated without enzyme for 72 h at 55°C (blank specimen), only traces of different sugars, i.e. glucose and raffinose, were detected. Note that because raffinose contains fructose, the very low quantities detected could have simply been reflecting impurities in the starch, rather than an actual degradation product. In comparison, when 1500 ppm enzyme was used to decompose the starch embedded in the agar gels, starch degradation products (such as glucose, raffinose, maltose and maltotriose) were observed, along with other unidentified oligosaccharides. Thus the embedded starch can be used, in principle, as an ingredient which can later be degraded; such a process could influence, and maybe even control, the porosity and mechanical properties of the sponge. Under the same experimental conditions (i.e. identical temperature, time, substrate and enzyme concentrations), enzyme solution decomposed a non-embedded starch solution with greater efficiency than a non-embedded counterpart, as evidenced by a chromatograph of degradation components (Gershon & Nussinovitch, 1998).

8.10.5 Hydrocolloid cellular solid as a carrier for vitamins

Vitamin A is an essential micro-nutrient involved in growth, epithelial maintenance, vision and reproduction (Sommer, 1994). Vitamin A deficiency (VAD) is a widespread problem, affecting mainly developing countries, which carries with it an increased risk of morbidity and mortality. Sub-clinical VAD has also been associated with high child morbidity and mortality, and vitamin A supplementation has been shown to reduce the statistics substantially (Rahmathullah *et al.*, 1990). Hydrocolloid sponges are in essence dry gel products. They were produced by preparing cold-set 1% alginate gels containing 1% soy oil, 500 ppm lecithin, sodium saccharin, 50 ppm β-carotene and vitamin A. These ingredients were homogenized and later incorporated into the gum solution, to which freshly prepared glucono-δ-lactone (GDL) solution was added. All gels were freeze-dried and kept over silica gel to avoid rehydration prior to testing, or were packaged in a laminate before clinical testing (Reifen *et al.*, 1998). The aim of the study was to evaluate a hydrocolloid carrier for vitamin A.

In the rural area of Gondar, Ethiopia, 220 families were screened for vitamin A status. Blood was obtained from 161 pre-school children, 2–5 years of age. Following the first blood drawing, 80 children were randomly chosen to receive the fortified edible sponges monthly for 3 months. A field worker visited the houses of the pre-school children every 2 weeks. Blood was drawn again after the 3-month period to determine vitamin A levels.

The average serum retinol level for the whole population was $0.74\,\mu\text{mol}\,\text{l}^{-1}$: 14 children (8.6%) were vitamin A deficient as defined by a serum retinol concentration of $<35\,\mu\text{mol}\,\text{l}^{-1}$ and 78 children (48%) had a low concentration ($<0.70\,\mu\text{mol}\,\text{l}^{-1}$). Serum retinol values for the study group which received the vitamin A-containing sponge for 3 months were as follows: $39 \pm 20.6\,\mu\text{mol}\,\text{l}^{-1}$ before intervention; $55.3 \pm 19.0\,\mu\text{mol}\,\text{l}^{-1}$ after 3 months of treatment. Vitamin A levels increased significantly following ingestion of the edible cellular solid, suggesting its usefulness as a carrier of vitamin A for children. In all cases but one, the levels of vitamin A in children who were below the normal range rose to normal following ingestion of the sponge. In one child, the level remained sub-normal despite sponge ingestion. Of the 80 children assigned to receive the edible sponges, three children refused to eat it. The other 77 ate the sponge all three times during the study period.

This study thus demonstrated the usefulness of edible sponges as a carrier for vitamin A. They could serve as a good means of supplementation or fortification for young children. The magnitude of the positive impact of supplementation may not have been seen in all children because of the short duration of the study. We assume that the longer the supplementation is maintained, the greater the opportunity for the children to restore their vitamin A levels. As mentioned, VAD is strongly linked to blindness, in addition to presenting an increased risk of morbidity and death from gastrointestinal and respiratory diseases. An edible sponge, which fulfills the criteria of a viable micro-nutrient carrier, could solve these problems by facilitating the delivery of a logical solution to this problem. However, more studies are needed with a larger population to test the ability of the carrier to include other micro-nutrients and vitamins before a general recommendation can be made. Since such sponges are designed for human consumption, coloring, flavoring and other ingredients can be added (to the slurry before gelation) to improve their potential acceptance. The concept of hydrocolloid sponges is new. There is a reasonable chance that these kinds of sponges will be used in the future not only as easily producible, affordable carriers for vitamins, but for other micro-elements as well, such as minerals and other food supplements.

8.10.6 Hydrocolloid cellular solids for water treatment

Water originating from food or agricultural/industrial processes can be contaminated with nitrate. Denitrification and its connection to immobilization is reviewed in this book (see Section 5.12 in Chapter 6) and elsewhere (Iza, 1991; Metcalf & Eddy, Inc., 1991; Power & Schepers, 1989; Tatsuo *et al.*, 1991). The aim here is to emphasize the importance of the cellular structures formed as a result of drying gels (composed of several ingredients) which serve as matrices for the immobilization of dentrifying isolates. One such isolate (*Pseudomonas* sp.) and starch were included in an alginate

(a)

(b)

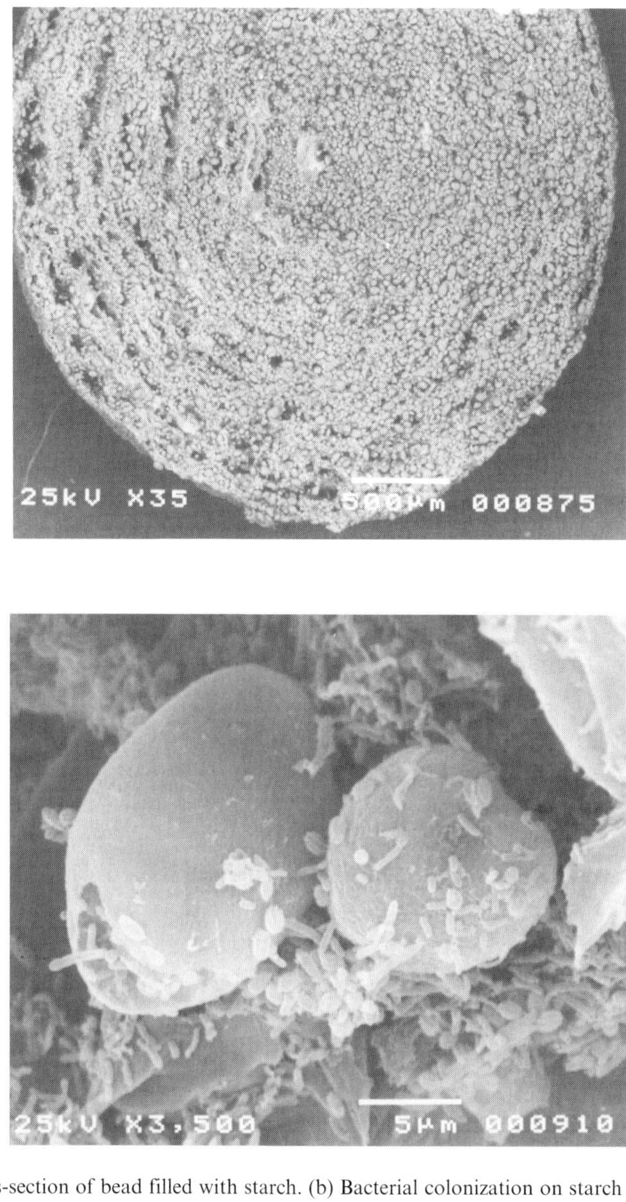

Fig. 8.6 (a) Cross-section of bead filled with starch. (b) Bacterial colonization on starch granules.

gel which was freeze-dried before incubation under denitrifying conditions. The dried beads had physical properties similar to those of porous, sponge-like matrices, with apparent air spaces within and around hydrocolloid-matrix walls. The beads revealed a matrix structure with variously sized pores, one that enabled gas release without matrix damage. The embodiment of starch granules within the matrix strengthened its structure. The dry matrices sustained their biological activity over a prolonged period. The drying that created the cellular solid did not damage the bacterial activity (Tal *et al.*, 1997). As stated, starch can be incorporated in freeze-dried alginate beads, serving as a carbon source and filler. Freeze-dried beads containing high concentrations of starch (Fig. 8.6) were found to have better mechanical and denitrifying properties than beads containing low concentrations of filler (Tal *et al.*, 1999).

8.10.7 Porosity control in cellular edible sponges

A method of controlling the porosity of cellular solid foods and their dependence on solid content was investigated. Freeze-dehydration of cold-set alginate gels was used to produce tailor-made porous solids. Gels were composed of sodium alginate, corn starch, oil and other additives required to induce gelation. Porosity values were derived from particle and bulk densities utilizing a helium pycnometer and a volumetric displacement method. To achieve a wider range of porosities, the gels were immersed in sucrose solutions of 10–60°Bx. Gel porosity after drying decreased from 0.85 to 0.42 and 0.07 after immersion in the 30 and 60°Bx solutions, respectively. Similarly, after immersing gels containing an additional 5% soy oil, porosity decreased to 0.36 and 0.04 after immersion in the 30 and 60°Bx solutions, respectively. Preparation and formulation play a significant role in porosity, and they can both be utilized to control its value within the range of 0.85–0.04. Scanning electron microscopy (SEM) micrographs revealed changes in the original cellular solid, characterized by numerous large void spaces, to a denser and more uniform appearance of the specimen after immersion in the 60°Bx sucrose solution. The dry gel system with controllable porosity provides a novel tool for the production of tailor-made cellular solid foods (Rassis *et al.*, 1997).

Edible cellular solids can be produced from alginate-starch gels immersed in a sucrose solution prior to their freeze-dehydration. This immersion resulted in an increase in the relative density of the dried samples and a decrease in their porosity. Closed cells were present especially at an intermediate time after the diffusion had begun and before termination. The mechanical properties were related to diffusion time (Rassis *et al.*, 1998). Another report discusses the inclusion of oil within such structures by emulsification in the gum solution before gelation occurs. Image analysis clearly showed encapsulated oil droplets of various areas within the produced cellular solid. Gaussian distribution of these areas was found with a maximum of 54% for the $(1-10)\times 10^{-12}\,m^2$ area range. The properties of the produced cellular solids were also dependent on relative humidity (RH) and storage conditions (Rassis *et al.*, 2000).

8.10.8 Dried gel beads

In this chapter, we describe the advantages of drying hydrocolloid gel beads in order to achieve higher strength, to have control over their porosity as a way for produced gases to be liberated from the matrix without interfering with its structure, and to produce desirable carriers for food and biotechnological operations. Agar gel spheres served as a study model for drying by desiccating air flow (air velocities 0.05–0.32 m s^{-1}), at temperatures slightly above and below the freezing point (from −9.3 to 10.7°C). A direct relationship between moisture vapor pressure at the sample surface and drying rates was observed (Suzuki *et al.*, 1989). Another report discusses the drying rate of agar gel spheres, at a wet-bulb temperature of 25°C and an air velocity of 3480 cm min^{-1}. Drying rate equations were developed, based on drying shell modes. The use of wet-bulb temperature was preferable to that of sample-center temperature (Kubota *et al.*, 1977). In addition to their many described applications and uses as models for drying studies, dry gel beads can be used for separation. Hydrogel (poly(*N*-isopropylacrylamide)) filtration was used to concentrate enzymes. After enzyme production by fermentation and centrifugation to remove cells, the supernatant was added to tubes containing the dry gel beads. Separation efficiency depended on temperature and was, at its highest value, 84% at 20°C (Jun-Han *et al.*, 1995). Another report of gel drying describes using plate temperatures of 70 and 90°C to dry agar gels, by radiation and conductive heat sources. Best results were obtained at 90°C, using conductive heat, which gave a more compact product (370% increase in product volume compared with a 450% increase with radiaton heating), while reducing the heating time by 20–50% (Maslyukov & Dement'eva, 1976). Stable gels composed of lard, sodium carboxymethylcellulose (CMC) and water were used to examine factors involved in the pro- and antioxidant activities of sodium chloride, other inorganic salts, heme compounds, meat fractions and other additives. The solid translucent gels, into which additives had been incorporated, were stored frozen or freeze-dried, or allowed to oxidize without physical change. When the gel was freeze-dried, a sponge-like structure was obtained which, after an induction period, auto-oxidized rapidly. Freezer-stored gels auto-oxidized at a rate roughly similar to freezer-stored meat (Ellis *et al.*, 1970).

8.10.9 Hydrocolloid composite cellular matrices

Composite materials are built from two or more components, each of which is made up of a different material and can be considered the matrix or particulate part of the system. The properties of the composite system are unique and are very rarely a simple average of those of the base components. There are many examples of composite materials in everyday use, for instance rubber wheels, fiberglass and different flexible packing materials. Many foods, such as filled chocolate, pastes, cheeses and cakes, are similar in composition to composite materials. Hydrocolloid cellular solids embedded with water-soluble polymer beads are another example of a particulate composite system. The physical properties of cellular solids built from a continuous matrix that contained spherical particles which differed in their composition, size and solid concentration were studied (Nussinovitch & Ungar, 1998; Ungar & Nussinovitch, 1998). Two model systems were tested. The first was built from an agar matrix (2%) that had embedded alginate beads, and the second consisted of a carrageenan (3%) matrix with chitosan beads. The quantity, sizes and concentration of the beads (fillers) embedded in the matrices were varied in order to study their influence on the physical properties of the composite systems and to determine their ability to control these properties.

The composite sponges were prepared by dripping alginate solution (1–2%) into a $CaCl_2$ solution (2%) in order to create spherical beads that were then placed in an agar solution (2%) that had been previously heated. The second model system was prepared by dripping chitosan solution (1–2%) into a sodium hexametaphosphate (SHMP) solution (2%). The resultant beads were introduced into a warm carrageenan solution (3%). The mixtures were then placed in a cool environment (4°C) to produce a gel system which, following freeze-dehydration, would become a dry cellular solid. Electron microscopy of the cellular materials was performed and the inner structure of the sponges was studied. Porosity (with a helium pycnometer) and mineral content analysis were conducted by an inductively coupled plasma (ICP) instrument. Several mechanical tests were carried out. The resultant composite sponges were compressed to ~80% deformation using an Instron Universal Testing Machine (UTM). The mechanical properties of the cellular composites were observed and then fitted into a compressibility model which describes the sigmoidal stress–strain relationship of cellular solids (Equation 7).

In the first stage of the research, same-size (2.5 mm) alginate and chitosan beads were introduced into the matrices and only their quantities differed. The maximum percentage of beads that could be introduced into the matrices was 80% in the agar–alginate system and 60% in the carrageenan–chitosan system. In both cases, the higher the percentage of beads introduced, the greater the strength of the dry composite material. Moreover the higher the proportion of beads within the dried matrix, the bigger their influence on the mechanical properties of the composite. From the prominence of the shoulder of the typical stress–strain curves and the values of the constant C_1, which is a scale factor of the stress, it was apparent that the composite systems had been strengthened by the addition of the spherical particles. This is due to the fact that alginate beads are a tougher material than the agar matrix. These beads are probably the reason for the rising values of Young's modulus (the parameter that describes strength of the system) with increasing percentage of beads. In the carrageenan–chitosan system, results differed because the chitosan beads collapsed during the compression tests. The chitosan particles formed during the compression tests and the compression of the matrix itself probably caused the strengthening of the system. The possibility of electrostatic interaction between the carrageenan matrix and the chitosan beads, and the adhesion that had existed during the gel stage, could have also contributed to the strengthening of the dry system (Nussinovitch & Ungar, 1998a, b).

During the second stage of the research, the effect of alginate and chitosan bead size on the mechanical properties of the agar–alginate and carrageenan–chitosan composite cellular solids was studied (Fig. 8.7). The diameter of the bead was increased from 2 to 5 mm. When the percentage of beads was 20%, the composite sponge systems were strengthened by the addition of larger embedded beads. Except for 5-mm alginate beads, the inclusion of 60% alginate or chitosan beads strengthened

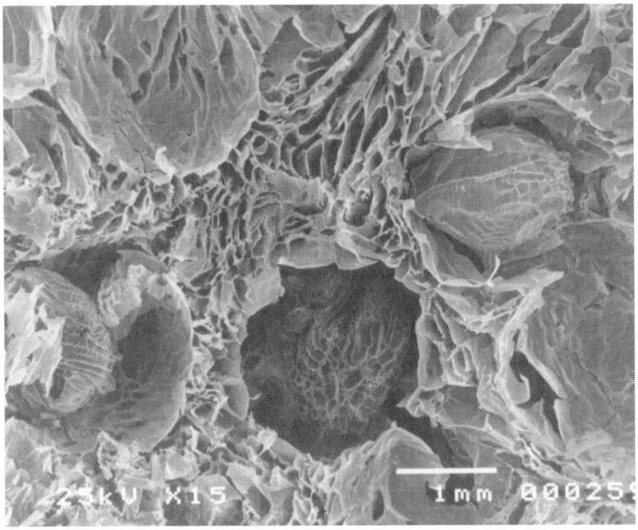

Fig. 8.7 SEM micrographs of dry agar matrix, including alginate beads embedded within its structure. (Courtesy of Y. Ungar.)

the cellular composite systems as well. In the third stage of the research, the effect of hydrocolloid concentration in the beads on the sponge composite systems was studied. Increasing bead hydrocolloid concentration from 1 to 2% strengthened the sponge systems. The Young's modulus values indicated that the composite systems were more rigid. The apparent strengthening of both the system with a small percentage (20%) of included beads and that with a high percentage (60%), indicated that the addition of dry cellular matter to the dry system probably causes the higher rigidity in the composite cellular systems. In mechanical testing conducted on the two sponge systems, a good match was found between stress–strain curves of the previously described compressibility model. The stress–strain curves were similar to those of other cellular solids and the correlation coefficient (R^2) was higher than 0.99. Electron microscopy of the sponges revealed voids around the beads embedded in the cellular matrix, due to bead shrinkage during the freeze-dehydration process. The size of the voids was dependent upon the level of bead shrinkage and upon differences in bead type, size and concentration. The alginate beads shrank to as much as 50% of their initial volume while the chitosan beads maintained their spherical shape and their level of shrinkage was far less (Nussinovitch & Ungar, 1998a, b).

Hydrocolloid concentration in the chitosan beads was an important factor because the 2% beads shrank to only 5% of their initial volume. The inclusion of beads within the matrix also enabled control over the porosities of the new cellular composite. When inclusion, on a wet weight basis, of 1% alginate in the 2% agar matrix was increased up to 50%, the porosity values were lower than those of the matrix. When the percentage of the inclusion was higher than 50%, i.e. 60 or 80%, the porosity increased to values more closely resembling those of the blank. The reason for this porosive phenomenon was the fact that the alginate bead shrank up to 50% during the freeze-dehydration process and voids around the beads were created. When the beads became the major part of the system, the voids around the beads became quite important and the sponges became very porous. The volumes of the voids around the beads were evaluated by computer analysis. With 40–80% inclusion, there was no difference in the space volume around a single bead. However, the number of voids changed with the percentage of included beads. From an estimate of the void volume, it was possible to calculate a theoretical system density by subtracting the void volume around the beads in the sponges, and a comparison was made with the actual experimentally determined density of the system. In the carrageenan–chitosan system, the porosity decreased with inclusion increases of up to 60%. In this system, the porosity values did not increase after 50% inclusion, probably because the chitosan beads shrank far less than the alginate beads and the voids around them were smaller. All the sponge specimens in this system gave high porosity values (~96%) because both the carrageenan matrix and the chitosan beads are porous materials and they both contribute to the porosity of the composite system (Nussinovitch & Ungar, 1998a).

A mathematical model describes the dependence of the modulus on the composite system. The model was fitted to describe a rigid system containing rigid particles in a rigid matrix. E and ϕ are the modulus and volume fraction of the particle p and matrix m, respectively:

$$E_c = x(E_p\phi_p + E_m\phi_m) + (1-x)(E_pE_m)/(E_p\phi_m + E_m\phi_p) \tag{14}$$

In the sponge specimens, there was no apparent interaction between the particles and the matrix and as a result, $x = 0$, and the expression becomes

$$E_c = E_pE_m/(E_p\phi_m + E_m\phi_p) \tag{15}$$

When the percent inclusion was less than 80, there was a correlation between Hirsh's modulus (defined by Equation 14), representing the theoretical model, and Young's modulus determined experimentally. In addition, a correlation was observed when alginate bead diameter was less than 4 mm. In summary, in this research, use of cellular solids in preparing composite sponge systems with included beads was demonstrated. The ability to control mechanical and porosity properties of the new cellular solid was gained by including the beads. The type of matrix and bead, quantity and size, and the interaction between the two components characterized the resultant physical properties. These types of composite systems can be used in the future for slow release, immobilization and packaging materials, or to produce new texture composite dry foods (Nussinovitch & Ungar, 1998b).

8.10.10 Acoustic properties of hydrocolloid cellular solids

Agar, κ-carrageenan and gellan gels (2%) with and without infused sucrose (5, 10 and 12.5%) were freeze-dried and the mechanical and acoustic signatures of the resulting solid sponges recorded (Fig. 8.8). For comparison, agar gels with 5 and 10% starch were also prepared and likewise tested. The presence of the sugar, or starch, in the dried gels increased their density in a manner that could not be predicted from the

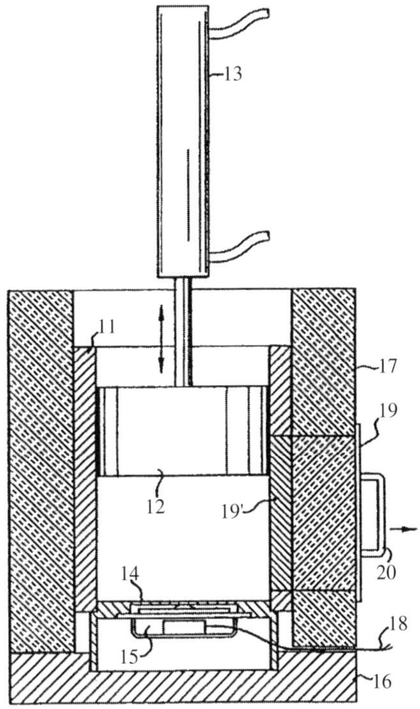

Fig. 8.8 Cell for compression and acoustic recording. (From Nussinovitch & Mey-Tal, 1998.)

corresponding stoichiometric relations between the gum and additive indicating that the sugar, or starch, were not inert fillers. In general, the presence of the sucrose in the dried gel solid matrix increased the brittleness of the sponges. The effect could be quantified in terms of the increase in the mechanical signature's apparent Richardson and Kolmogorov fractal dimension (Nussinovitch & Mey-Tal, 1998; Nussinovitch *et al.*, 2001). The more brittle dried gels had a *richer* acoustic signature (Fig. 8.9) whose apparent fractal dimension was determined with the *blanket* algorithm

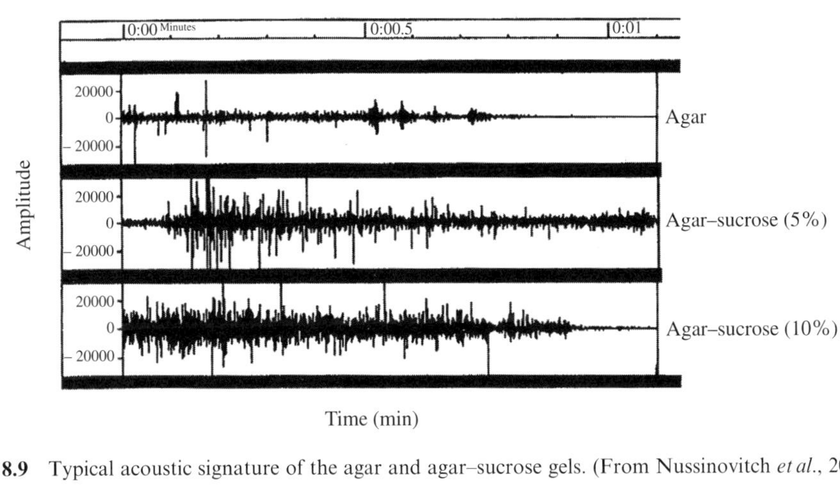

Fig. 8.9 Typical acoustic signature of the agar and agar–sucrose gels. (From Nussinovitch *et al.*, 2000.)

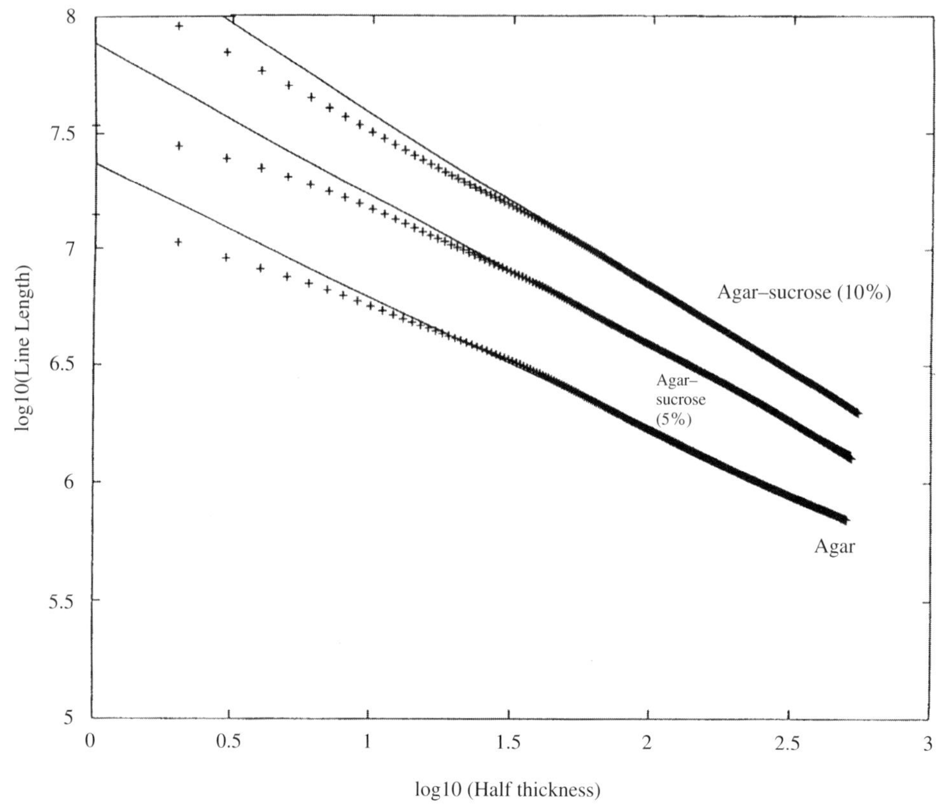

Fig. 8.10 Typical Richardson's plot of the acoustic signature of the dried agar and agar–sucrose gels created with the *blanket* algorithm. (From Nussinovitch *et al.*, 2000.)

(Fig. 8.10). Despite the irregular and jagged appearance of all the dry gels' stress–strain relationships, they could be described by the same kind of a three-parameter empirical model originally derived for soft baked products and polymeric foams. Although the added sugar, or starch, stiffened the dehydrated gels the effect could not be quantified unambiguously because of the differences in the density and in the overall shape of the stress–strain curve (Nussinovitch *et al.*, 2000).

8.11 References

Attenburrow, G.E., R.M. Goodband, L.J. Taylor *et al.* (1989) Structure, mechanics and texture of a food sponge. *J. Cereal Sci.*, **9**, 61–70.

Barret, A. & M. Peleg (1992) Relationship between extrudate cell structure and texture. *J. Food Sci.*, **57**, 1235–1257.

Barret, A.M., M.D. Normand, M. Peleg & E. Rose (1992) Characterization of the jagged stress–strain relationships of puffed extrudates using the Fast Fourier Transform and Fracta Analysis. *J. Food Sci.*, **57**, 227–232, 235.

Ellis, R., A.M. Gaddis, G.T. Currie & F.E. Thornton (1970) Sodium chloride effect on autoxidation of the lard component of a gel. *J. Food Sci.*, **35**, 52–56.

Gershon, Z. & A. Nussinovitch (1998) Characteristics of enzymatically produced agar–starch sponges. *Food Hydrocolloids*, **12**, 105–110.

Gibson, L.J. & M.F. Ashby (1988) *Cellular Solids Structure and Properties*. Pergamon Press, Oxford.

Harris, M. & M. Peleg (1996) Patterns of textural changes in brittle cellular cereal foods caused by moisture sorption. *Cereal Chem.*, **73**, 225–231.

Iza, J. (1991) Fluidized bed reactors for anaerobic wastewater treatment. *Water Sci. Technol.*, **24**, 109–132.

Jeronomidis, G. (1988) Structure and properties of liquid and solid foams. In: (Blanshard, J.M.V. & J.R. Mitchell, Eds.), *Food Structure*. Butterworth, pp. 59–75.

Jun Han, Chang Ho Park & R. Ruan (1995) Concentrating alkaline serine protease, subtilisin, using a temperature sensitive hydrogel. *Biotechnol. Lett.*, **17**(8), 851–852.

Kaletunc, G., M.D. Normand, E.A. Johnson & M. Peleg (1991) Degree of elasticity determination in solid foods. *J. Food Sci.*, **56**, 950–953.

Kaletunc, G., M.D. Normand, E.A. Johnson & M. Peleg (1992) Instrumental determination of elasticity of marshmallow. *J. Texture Studies*, **23**, 47–56.

Kubota, K., H. Kobatake, K. Suzuki & H. Hosaka (1977) Drying rate equations of agar gel and carrot based on drying-shell model. *J. Faculty of Fisheries and Animal Husbandry*, **16**(2), 123–130.

Lillford, P.J. (1989) Structure and properties of solid food foams. In: (Wiloson, A.J., Ed.), *Foams, Physics, Chemistry and Structure*. Springer Verlag, London, pp. 149–166.

Maslyukov, Yu. P. & N.A. Dement'eva (1976) Freeze drying of agar gel using radiation and conductive heat sources. *Issledovaniya-po-Tekjnologii Rybnukh Produktov*, **6**, 129–131, 153–154.

Meincecke, E.A. & R.C. Clark (Eds.) (1973) *Mechanical Properties of Polymeric Foams*. Technomic, Westport, CT.

Metcalf and Eddy, Inc., (1991) *Wastewater Engineering Treatment, Disposal, Reuse*, 3rd edn. McGraw-Hill International Editions: New York, p. 1334.

Nejedly, B., J. Hanzl & J. Kodet (1986) Maltodextrin sponge based on starch hydrolysate and egg. Czechoslovakian Patent CS 247434.

Nuebel, C. & M. Peleg (1993) Compressive stress-strain relationships of two puffed cereals in bulk. *J. Food Sci.*, **58**, 1356–1360, 1374.

Nuebel, C. & M. Peleg (1994) Compressive stress-strain relationships of agglomerated instant coffee. *J. Food Process Eng.*, **17**, 383–400.

Nussinovitch, A. (1995) Compressive characteristics of hydrocolloid Sponges. In: *The 8th International Conference and Industrial Exhibition on Gums and Stabilisers for the Food Industry, 1–14 July*. The North East Wales Institute, Cartrefle College, Wrexham, CLWYD, UK.

Nussinovitch, A. (1997) In: *Hydrocolloid Applications, Gum Technology in the Food and Other Industries*. Blackie Academic & Professional, London, pp. 292–311.

Nussinovitch, A. & Z. Gershon (1997a) Alginate-oil sponges. *Food Hydrocolloids*, **11**(3), 281–286.

Nussinovitch, A. & Z. Gershon (1997b) Physical characteristics of agar-yeast sponges. *Food Hydrocolloids*, **11**(2), 231–237.

Nussinovitch, A. & E. Mey-Tal (1998) System for measuring the crispiness of materials. United States Patent 5,827,974.

Nussinovitch, A. & Y. Ungar (1998a) Cellular dry macro-capsules. Provisional application # 60/111,643 from 09/12/1988.

Nussinovitch, A. & Y.Ungar (1998b) New cellular-dry carriers. United States application date 09/12/98; Provisional application # 60/111,645

Nussinovitch, A., M. Peleg & M.D. Normand (1989) A modified Maxwell and a non-exponential model for characterization of the stress relaxation of agar and alginate gels. *J. Food Sci.*, **54**, 1013–1016.

Nussinovitch, A., G. Cohen & M. Peleg (1991) Comparison of the compressive characteristics of puffed popcorn and polystyrene foam particles. *J. Cellular Plastics*, **27**, 527–539.

Nussinovitch, A., M. Steffens, P. Chinachoti & M. Peleg (1992a) Effect of strain level and storage time on the recoverable work of compressed bread crumb. *J. Texture Studies*, **23**, 13–24.

Nussinovitch, A., R. Velez-Silvestre & M. Peleg (1992b) Mechanical properties of hydrocolloids gels filled with internally produced CO_2 gas bubbles. *Biotechnol. Prog.*, **8**, 424–428.

Nussinovitch, A., R. Velez-Silvestre & M. Peleg (1993) Compressive characteristics of freeze-dried agar and alginate gel sponges. *Biotechnol. Prog.*, **9**, 101–104.

Nussinovitch, A., M. Nussinovitch, R. Shapira & Z. Gershon (1994) Influence of immobilization of bacteria, yeasts and fungi spores on the mechanical properties of agar and alginate gels. *Food Hydrocolloids*, **8**(3–4), 361–372.

Nussinovitch, A., L. Peleg & Z. Gershon (1995) Properties of agar-starch sponges. *The 8th International Conference and Industrial Exhibition on Gums and Stabilsers for the Food Industry, 1–14 July*. The North East Wales Institute, Cartrefle College, Wrexham, CLWYD, UK.

Nussinovitch, A., Z. Gershon & L. Peleg (1998) Characteristics of enzymatically produced agar–starch sponges. *Food Hydrocolloids*, **12**(1), 105–110.

Nussinovitch, A., M.G. Corradini, M.D. Normand & M. Peleg (2000) Effect of sucrose on the mechanical and acoustic properties of freeze-dried agar, κ-carrageenan and gellan gels. *J. Texture Studies*, **31**, 205–223.

Nussinovitch, A., M.G. Corradini, M.D. Normand & M. Peleg (2001) Effect of starch, sucrose and their combinations on the mechanical and acoustic properties of freeze-dried alginate gels. *Food Research International*, **34**, 871–878.

Peleg, M. (1993) Calculation of the compressive stress–strain relationships of layered arrays of cellular solids using equation solving computer software. *J. Cellular Plastics*, **29**, 285–293.

Peleg, M. (1997) Review: Mechanical properties of dry cellular solid foods. *Food Science and Technology International*, **3**, 227–240.

Peleg, M., I. Roy, O.H. Campanella & M.D. Normand (1989) Mathematical characterization of the compressive stress–strain relationships of spongy baked goods. *J. Food Sci.*, **54**, 947–949.

Phillips, P.J. & N.R. Waterman (1974) The mechanical properties of high-density rigid polyurethane foams in compression. I. Modulus. *Polymer Eng. Sci.*, **14**, 67–70.

Power, J.F. & J.S. Schepers (1989) Nitrate contamination of ground-water in North America. *Agric. Ecosys. Environ.*, **26**, 165–187.

Rahmathullah, L., B.A. Underwood, R.D. Thulasiraj, R.C. Milton, R. Ramaswamy, G. Rahmathullah & G. Babu (1990) Reduced mortality among children in southern India receiving small weekly dose of vitamin A. *N. Engl. J. Med.*, **323**, 929–935.

Rassis, D., A. Nussinovitch & I.S. Saguy (1997) Tailor-made porous solid foods. *International J. Food Sci. Technol.*, **32**(4), 271–278.

Rassis, D.K., I.A. Saguy & A. Nussinovitch (1998) Physical properties of alginate–starch cellular sponges. *J. Agric. Food Chem.*, **46**(8), 2981–2987.

Rassis, D., A. Nussinovitch & I.S. Saguy (2000) Oxidation of encapsulated oil in tailor-made cellular solid. *J. Agric. Food Chem.*, **48**, 1483–1489.

Reifen, R., M. Edris & A. Nussinovitch (1998) A novel, vitamin A fortified, edible hydrocolloid sponge for children. *Food Hydrocolloids*, **12**(1), 111–114.

Sommer, A. (1994) Vitamin A: its effect on childhood sight and life. *Scand. J. Nutr., Suppl.*, **27**, 60–66; 71.

Suzuki, K., H. Onishi, M. Esaka, K. Kubota & H. Hosaka (1989) Low temperature drying of agar gel sphere in desiccated cold air flow. *J. Food Sci.*, **54**(2), 416–418.

Swyngedau, S., A. Nussinovitch & M. Peleg (1991a) Models for the compressibility of layered polymeric sponges. *Polymer Engineering and Science*, **31**, 140–144.

Swyngedau, S., A. Nussinovitch, I. Roy, M. Peleg & V. Huang (1991b) Comparison of four models for the compressibility of breads and plastic foams. *J. Food Sci.*, **56**, 756–759.

Swyngedau, S. & M. Peleg (1992) Characterization and prediction of the stress–strain relationship of layered arrays of spongy baked goods. *Cereal Chemistry*, **69**, 217–221.

Tal, Y., J. van Rijn & A. Nussinovitch (1997) Improvement of structural and mechanical properties of denitrifying alginate beads by freeze-drying. *Biotechnol. Prog.*, **13**, 788–793.

Tal, Y., J. van Rijn & A. Nussinovitch (1999) Improvement of mechanical and biological properties of freeze-dried denitrifying alginate beads by using starch as a filler and carbon source. *Appl. Microbiol. Biotechnol.*, **51**, 773–779.

Tatsuo, S., N. Hiroki & M. Naomichi (1991) Immobilization of activated sludge by the acrylamide method. *J. Fermentation and Bioengineering*, **2**, 141–143.

Torres, A.Q., H.G. Schwartzberg, M. Peleg & R. Rufner (1978) Textural properties of amylose sponges. *J. Food Sci.*, **43**, 1006–1010.

Ungar, Y. & A. Nussinovitch (1998) Composite cellular carriers. *Bioencapsulation VII and Mini-Symposium on Microencapsulation*. Tidewater Inn, November 20–23. Easton, Maryland, USA.

Warburton, S.C., A.M. Donald & A.C. Smith (1990) The deformation of brittle starch foams. *J. Materials Sci.*, **25**, 4001–4007.

Wollny, M. & M. Peleg (1994) A model of moisture induced plasticization of crunchy snacks based on Fermi's distribution function. *J. Food Sci. Agric.*, **64**, 467–473.

Chapter 9
Hydrocolloids in the Production of Special Textures

9.1 Introduction

Gum constituents are a part of almost all natural foods, contributing to the structure and textural properties of the plant (Howes, 1949). When gums are used as food additives they impart functional and textural properties to the finished product. Gums can change the processing parameters of foods and influence water retention, evaporation and freezing rates, the formation of ice crystals and participation in chemical reactions. Hydrocolloids have a thickening capability upon dispersion in water media, and they are thus used as bodying, stabilizing and emulsifying agents in foods such as pie fillings, dry beverage mixes, soups, sauces and sauce mixes. Other gums also have gelling properties and are used to construct products that will not flow without applied pressure and thus retain their shape. Good examples are pectin-gelled cranberry sauce and starch-based milk puddings (Nussinovitch, 1997).

9.2 Fabricated foods

Using fish, meat, fruit or vegetables as main ingredients within a matrix, which is usually produced from a gum, can create fabricated foods. Since both matrix builders and embedded ingredients are numerous, there are endless possibilities for creating these types of foods which serve as supplies, for the introduction of new textures and to reduce costs in developing societies (Aguilera & Stanley, 1986). Gums were incorporated into meat products to achieve better control of their texture, improve sliceability and increase yield. In some meat products, hydrocolloids are responsible for the undesirable broad dark striations (called tiger stripping), running parallel to the meat fibers (Philp *et al.*, 1998). The swelling ability of the type of carrageenan used influences its activity within the product. Semi-refined carrageenan (less swelling) improved performance in injected poultry by reducing the incidence of tiger stripping without reducing purge controls (Philp *et al.*, 1998).

Fabricated fruit is easily manufactured with alginates. A gel is readily formed when a soluble calcium salt is added to a sodium alginate solution. This gel is stable over a wide range of temperatures, has excellent syneresis control, and is irreversible to heat. Possible uses for this fabrication concept include imitation cocktail cherries; imitation glazed fruit pieces for cakes, breads, cookies, ice cream and candy products; icing; and gelled products containing pureed fruit (Fig. 9.1); also see chapter 7 on fruit texturization (Aguilera, 1992; Ben-Zion & Nussinovitch, 1996; Glicksman, 1976; Kaletunc *et al.*, 1989; Lodge, 1981; Luh *et al.*, 1976; Mitchell & Blanshard, 1976; Nussinovitch & Peleg, 1990; Nussinovitch *et al.*, 1991a, b; Ring & Stainsby, 1982; Silberberg, 1989; Szczesniak, 1968; Tolstoguzov, 1971; Tolstoguzov & Braudo, 1983; Unilever Ltd., 1974; Weiner & Nussinovitch, 1994; Wood, 1975). With variations on the technique, it is possible to produce vegetable products and spaghetti, as well as exotic foods, such as imitation caviar (Davidson, 1980). Fabricated onion rings were introduced in the 1970s. A slurry, consisting mainly of sodium alginate and diced onions, is extruded into a calcium chloride bath. After the skin forms, the rings are dusted with breading batter, fried in deep fat, blast-frozen and packed. Cellulose ethers, which form lipophobic protective films, are used to prevent oil absorption during frying. Other hydrocolloids – carrageenan, gelatin (and recently gellan), and combinations of gums such as carrageenan and locust bean gum (LBG) – have been used to fabricate food products. Examples include reconstituted pimento strips (based on alginate and gum arabic), the aforementioned imitation caviar, and restructured fish and shellfish (Nussinovitch, 1993, 1997).

Use of sodium carboxymethyl cellulose (CMC) in food applications is on the rise, especially in developed countries where the popularity and convenience of fabricated

Fig. 9.1 Fabricated vegetables and fruits.

foods has grown rapidly since the early 1950s. Cellulose gum is used in such foods to improve product acceptability (Davidson, 1980). Hydroxypropylcellulose can also be used in fabricated foods to a large extent. Its useful properties are its ability to form solvent-soluble films and its surface-active stabilization (Davidson, 1980).

During processing, the temperature at a given point within a fabricated food depends on the rate of heating or cooling and the position in the food of the point being measured. The temperature therefore changes continuously. Factors that influence temperature include the temperature of the heating medium, the thermal conductivity of the fabricated food and its specific heat (Fellows, 2000). Thermal diffusivity is calculated by dividing the thermal conductivity by the product of density multiplied by the specific heat capacity. To calculate non-steady-state heat transfer in a single direction, the change of temperature with time needs to be determined (Fellows, 2000). Theoretical formulas for the prediction of effective thermal diffusivity on one-dimensional non-steady-state heat conduction in non-frozen and frozen fish and meats were investigated. Theoretical results agreed well with measured values. The effective thermal diffusivity for fabricated foods was predicted by preparing an appropriate ternary contour diagram (Kong *et al.*, 1994).

Extrusion processes for the manufacture of snacks use a flour or starch matrix combined with flavors, colors and spices. Heat and pressure to yield the desired functional properties in the finished product transform the starch component. Hydrocolloids such as cellulose ethers are used to improve thermostability and to provide lubrication during extrusion. Starches are also widely used in conventional batch-canning operations to maintain the consistency of the product after heat treatment. Such starches need to be stable over a wide range of storage conditions and must retain their physical and sensory qualities. This requirement is especially important for creamed vegetable soups, gravy-based products and pie fillings. Modified starches that do not thin out under continuous heating and do not gel when stored at low temperatures are used for this purpose (Nussinovitch, 1993, 1997).

9.3 Japanese desserts and molded products

Agar has a wide variety of uses. It is employed in the bakery, confectionery and dairy industries. Agar stabilizes icing and prevents adhesion of the sugar coating to the wrapper. It can be used as a doughnut glaze stabilizer. By increasing the viscosity of the glaze, the stabilizer increases its adherence to the doughnut and provides quicker setting and flexibility with reduced chipping and cracking. Agar acts as a bulking agent and is useful in low-calorie breads or biscuits due to its non-nutritive nature. Besides all of these applications, agar makes a special contribution to the sweet dessert industry. Agar in the form of a sweetened and flavored gel has been an important item in the Asian diet for several hundred years. Mitsumame is a sweet dessert based on agar gel, which consists of an attractive, colorful product usually served in glass bowls. The dessert is prepared by soaking 237 ml of 1% agar gel cubes ($\sim$1 cm to a side) in syrup, together with pieces of cherries, pineapple, peas (*mame* in mitsumame means peas), and sweetened mandarin orange. The uniqueness of agar in this application is its ability to maintain the straight edges of the cubes, something which cannot be done with other gels, and to permit heat treatment of such a canned product, provided that the selected agar has a high melting point (minimum 85°C).

Yokan is a sweet, jelly-like confection, which is very popular in Japan. The confection is prepared from agar, mashed Azuki beans and sugar, and is basically a texturized vegetable. Instead of the Azuki beans, powdered green tea, brown sugar, honey, and the juice of yuzu rind, grapes, red plums or jams such as apricot, apple or persimmon can be used. Chestnuts can also be used to replace the beans, but they are very expensive. Yokan is composed of 26.5% water, 54.3% sugar, 0.2% salt, 18% (dry weight) mashed beans and 1% agar. Yokans are molded into ingot-shaped bars, and are generally sealed and packaged in laminated film containers. A modernized version of this traditional Japanese confection is a condensed sweet agar jelly, which is extremely sticky. It is usually wrapped in an edible agar-based film called *oblate*. To produce the jelly, 50 g agar, $1–2 \times 10^3$ g sugar, $1.5–3.0 \times 10^3$ g invert sugar, 2 g citric acid and $2.0–3.0 \times 10^3$ g water are used, and color and flavor are added as needed. To produce the jelly, agar is dissolved in boiling water, to which the sugars are then added. The viscous solution is heated to 106°C. During this time, and as a result of water evaporation, the solution condenses. After brief cooling, fruit juice and/or synthetic organic acid, flavoring agents and color additives are added to the mixture which is allowed to gel at room temperature overnight. The solid content after this process is ~70–75%. The product is then dried at 55°C for 15–30 h to reduce the moisture to 16–18%, giving a solid content of ~83% in the final product. The dried product is cut into small pieces, which are manually wrapped, first in *oblate*, made from starch and agar, then in cellophane, for a film thickness of ~10–15 μm. It is usually sold in drugstores in Japan in 9.3-cm-diameter circles. Adding a 5% aqueous starch solution to a 2.5% aqueous agar solution at a ratio of 1:2 produces the film. The hot mixed solution is brushed onto the glassy surface of a metal or glass body and the final product after drying is a rigid dry film that can be removed and cut to a predetermined shape. This thin film can be used to wrap condensed sweet agar jelly or other sticky materials, as well as for dosing powdered medicines.

The aforementioned Japanese desserts are regularly composed of one layer. The so-called simultaneous method can be used to produce similar canned dessert gels; however these gels contain more than one layer to change their look and aesthetic appeal. These products are prepared from commercially produced gums. For example solutions of a blend of xanthan gum/LBG and carrageenan are poured into a mold at 158°F (70°C) and cooled, whereupon they separate into two distinct layers. These products can contain juice, coffee extract or dietary fiber, as well as different colorings, flavorings and texturizing materials (Nussinovitch, 1993).

9.4 Use of hydrocolloids in spray-dried products

Spray-drying is frequently used for producing fixed flavors. During the drying process, water is removed and the flavoring is coated with a film of gum that protects it against oxidation or other changes. An oil-in-water emulsion is made with a suitable hydrocolloid that encapsulates the flavor droplets when the water is removed, preventing their evaporation, oxidative deterioration or moisture absorption from the air; the better the emulsion, the better the protection for the dried particles (Glicksman, 1969). Extensive use has been found for gum arabic as a flavor fixative. Spray-drying is a fairly simple production operation relying on the proper selection of spray-dryer nozzles and inlet pressures, and different drying times. Gum arabic as a flavor fixative forms a thin film around the flavor particle, protecting it from oxidation and evaporation and preventing it from absorbing moisture from the air. Gum arabic is best suited to this purpose, but ghatti gum, another plant exudate, is also used. Gum arabic reduces surface tension, enabling the emulsion to form more easily. It also coats dispersed oil droplets so that they are surrounded by an electrically charged surface. Since all the droplets then have the same charge, they repel each other, inhibiting collision and coalescence. Gum arabic is widely used for micro-encapsulation. Modified starches have been used for the same purpose when gum arabic is unavailable, but with less success (Nussinovitch, 1997).

Tests on spray-dried emulsions of easily oxidized aldehydes showed that the film protects the products from oxidation for years, whereas unprotected materials oxidize in seconds. Citrus oils and imitation flavors are prepared by emulsifying them in gum arabic solution before spray-drying them. The particle size can range from 1 to 50 μm. Other colloids are also used for these purposes, but gum arabic is the most important and popular (Davidson, 1980). Some beverage flavor emulsions make use of the additive properties of gum arabic and gum tragacanth. In such cases, gum tragacanth is the emulsion stabilizer and gum arabic provides smoothness (Davidson, 1980). Gum ghatti can be used as a stabilizer for oil-in-water emulsions, replacing gum arabic as the emulsifiying agent. It is used in oil-soluble vitamin products and as a stabilizer in maple syrups containing butter for use on pancakes and waffles (Topalian & Elsesser, 1966).

The goal in food drying is to remove water at the lowest possible temperature, thereby optimizing product quality. With fruits and vegetables, this sometimes results in a tough, leathery surface on the dried particle, which can adversely affect rehydration and eating qualities. The problem can be reduced by wetting the fruit or vegetable particles with cold solutions of water-soluble gums such as guar or CMC prior to drying. Several methods are used, for example, soaking under vacuum to improve penetration of the hydrocolloid into the particles (Nussinovitch, 1997). In dry instant beverages, cellulose gum improves oral feel, stabilizes the system and contributes to the suspension of pulp solids and to the overall body of the product. In dehydrated foods, cellulose gums facilitate hydration. Cellulose gum improves baked goods in the sense that they retain moisture better and exhibit improved keeping quality.

9.5 Frozen products

Frozen desserts are mixtures of ice crystals in flavored liquid syrup. The most common frozen dessert is ice cream. During the last 50 years, a huge change in the texture of ice-cream products has occurred, from home-made ice cream that suffered from sandiness, lack of smooth meltdown and presence of ice crystals to ice creams with none of these defects as a result of the addition of various hydrocolloids (Glicksman, 1969).

Gum karaya can be used as a stabilizer in ice cream, ice milk, mellorine and related products. In ice pops and sherbets, formation of large ice crystals and syneresis can be prevented by including 0.2–0.4% gum karaya. Combinations of 0.15% gum karaya and 0.15% LBG can be used successfully for ice pop stabilization. In meat products such as bologna, ~0.25% karaya provides cohesiveness and good water-retention properties. Karaya, as well as carrageenan, can be used as a binder and emulsifying agent in quantities of less than ~1%. The binders are used to absorb the water resulting from the ice during chopping (Glicksman, 1969). LBG is used in the food industry for its ability to bind and immobilize large amounts of water. This property helps inhibit ice crystal formation in frozen products, produce viscosity, modify texture and stabilize product consistency in the face of temperature changes (Glicksman, 1969). Sodium CMC in its highest purified form is used in many food applications. In frozen desserts (such as ice cream), cellulose gum inhibits the formation of ice crystals (Davidson, 1980). Gum arabic, because of its water-absorbing properties, is frequently used as a stabilizer in frozen products such as sherbets, ices and ice creams. Gum inclusion inhibits the formation and growth of ice crystals. Other stabilizers, such as carrageenan and LBG, can be used for the same purposes (Glicksman, 1969). Guar gum is used in the food industry for its ability to bond and immobilize large amounts of water. This property contributes to inhibition of ice crystal formation, product texture, stabilization of product consistency to changes in temperature, and viscosity (Davidson, 1980).

Freezing often causes undesirable changes in foods, and hydrocolloids are used to improve their quality. To produce a high-quality ice cream, a blend of guar gum or CMC with a smaller amount of carrageenan may be used. If xanthan and guar gum are used instead, viscosity is lower and faster processing is obtained. Karaya gum has been used in the past as a stabilizer for frozen desserts, but has been replaced almost completely by other gums. Carrageenan, guar gum and CMC have also been used as stabilizers in other frozen products. In foods containing starch as the main ingredient, there is a tendency for water to exude from the gel. Thus starch-based products curdle and undergo syneresis (loss of water) after freezing and thawing. Modified starches have been developed to deal with the problem (Nussinovitch, 1993, 1997). Gum karaya can be used in ice pops, sherbets, cheese spread, ground meat preparation and meringue products. Sherbets that contain no less than 1% by weight of milk fat and less than 4% by weight of total milk solids are stabilized by a combination of stabilizers and emulsifiers at no more than 0.5% of the total mix (Davidson, 1980).

Freeze-texturization was used for the production of porous starch-based sponges (Torres, 1977). Sponges prepared from high-amylose cornstarch had a *fibrillar crab-meat-like* character, whereas those prepared from pure amylose were *flaky and meat-like*. Sponges consist largely of gelatinized starch and thus possess high water-holding capacity and low solubility (Nussinovitch, 1997). Freezing is also used to modify the texture of sponge-like products produced from soybean curd. In freeze-texturization, cooling below the solution's freezing point initially forms two phases (ice and concentrated solution/suspension). Thus a fibrillar structure is induced by the controlled growth and orientation of the ice phase favored by unidirectional freezing, followed by fixation via protein–protein interactions (Lugay & Kim, 1981). Lillford (1985) discussed the physicochemical principles involved in the freeze-texturization process. The rate at which freezing is induced determines the character of the formed structure (spongy or fibrous). Flow-freezing to achieve a crystal growth rate which exceeds that of nucleation is necessary to achieve a homogeneous texture. Mechanical means are also used to break ice crystals, to form fibers and to induce alignment (Nussinovitch, 1997).

Frozen doughs are widely used in industrial bakeries to make baking more profitable. Transportation is facilitated and labor costs are reduced as fresh breads are baked in bake-off stations when needed by consumers. However, loaf volumes are usually smaller and quality poorer for breads baked from frozen doughs, especially in those with low fat content (Rasanen & Autio, 1998). Addition of hydrocolloids such as CMC, alginate, and different blends in quantities of up to 1.5% yielded higher total dough water content without changing baking properties. There were no obvious differences between analyzed samples or added hydrocolloid levels. Thus, the baking properties are related to the correct balance between dough elasticity, amount of *free* water and gassing power after frozen storage (Rasanen & Autio, 1998).

9.6 Stabilization and stable foams

Many milk and juice products are processed aseptically, and specialized carrageenans are used in such products. These carrageenans are soluble at ultra-high temperatures, stabilize the milk protein, suspend fat globules, and preserve the product's taste; they also permit hot or cold packaging of the finished product. Carrageenan is also widely used in other milk products, such as processed beverages, pasteurized milk, dry-mix beverage powders, milk puddings and cheese products (Nussinovitch, 1997). Gelation and thickening of dairy products is one of the most important food applications of the carrageenans, particularly κ-carrageenan. Low-methoxy pectins (LMP) also form gels with milk and these too have food applications mainly in dairy desserts. Rheometry and differential scanning calorimetry have been used to study the mixed gels formed with skimmed milk powder, κ-carrageenan and LMP. Very different gelation mechanisms have been observed for the two polysaccharides. κ-Carrageenan forms a complex with casein micelles and it appears to create a molecular *Velcro* interaction, with the free ends of bound κ-carrageenan molecules linking casein micelles to form a gel network. At high ratios of κ-carrageenan to skimmed milk powder, a pure κ-carrageenan network also appears to form, presumably consisting of more extended cross-linked κ-carrageenan structures within which the casein micelles are enmeshed. In contrast, there is no direct interaction between casein micelles and LMP. The casein appears simply to act as a source of calcium ions which promote gelation of the pectin (Oakenfull *et al.*, 2000). Carrageenan is also used in the food industry for gelation, thickening, stabilization, and improvement of palatability and appearance. In dry mixes, e.g. milkshake and instant breakfast powders, carrageenan is used to suspend the ingredients and to impart richness and body to these drinks. In manufactured products, carrageenan is used to keep ingredients in suspension and to give the product a rich mouth feel. Many other milk applications can be found elsewhere. Typical water applications of carrageenan include its use in dessert gels, low-calorie jellies, fish gels, syrups, fruit drink powders and frozen concentrates, relishes, pizza, barbecue sauces, imitation milk, imitation coffee creams, whipped toppings, puddings, air-treatment gels, toothpastes and lotions (Davidson, 1980).

The formation of stable foams is essential in preparing many food products, ranging from bread to whipped cream and mousses. Air is incorporated to produce the desired texture in the final product, a porous, thin-walled structure. Starches, and sometimes proteins, can be induced to foam via two main mechanisms: sudden vaporization of a solvent, or generation of a gaseous phase in proofing and dough baking. Techniques such as vapor-induced puffing and explosion puffing are widespread. The texture of solid starch-based foams depends on the amylose:amylopectin ratio. High-amylose products are hard and dense, whereas high-amylopectin products are fragile, but can be modified by amylose supplementation to yield a structure which is more resistant to breakage (Matz, 1984).

Gums, frequently gum mixtures – such as carrageenan, sodium alginate, guar and CMC – are used to stabilize these foams. Cellulose ethers are used to stabilize whipped toppings. Although protein-based foams can be stabilized by physical means (for example, heating or cooling), gums such as carrageenan, alginate and LBG are often added to react with the protein and form a stable foam (Nussinovitch, 1993). Agar is often used in marshmallow stabilization and gum arabic is added to soften texture and produce a more desirable product (Werbin, 1960). Pectin can help stabilize aerated confectionery products.

The effects of sugars on the stability of egg foams were investigated using a rheological technique, pulsed nuclear magnetic resonance (NMR) measurements and some other tests. The linear viscoelastic region for foam systems under uniaxial compression was very narrow. The creep behavior for foam systems could be divided into three stages: elastic, viscoelastic and steady flow. The creep compliance and steady-flow stage of the foam systems decreased with increasing sugar concentration. The amounts of drainage from the foam systems increased, and protein contents in the drainage liquid decreased with standing time. The rate of drainage was impeded when sucrose was added to the albumen. The foam systems became more dense upon addition of saccharides, because the particle size of the gas cells become smaller and more homogeneous in the presence of sugar. Moreover, the spin–spin relaxation times of foam systems became shorter upon addition of sugar. Saccharides may be thought to prevent the mobility of water between films of foam, thereby stabilizing the foam systems (Ochi *et al.*, 2000).

9.7 Candies

Candies are a popular product among children and adults and their versatility is visually alluring as well as pleasing to the consumer. The confectionery industry uses gum arabic to a great extent, for crystallization prevention, as an emulsification agent of fat and as a glaze in candies, chewing gum and cough drops. Gum arabic serves to coat the center of sugar-coated tablets. It is the main ingredient in gumdrops (regular and dietetic) and other chewy-type gums, where pectin or modified starches can also play a major role (Davidson, 1980). The incorporation of sorbitol, mannitol and gum arabic can produce dietetic candies. The higher the gum arabic content, the softer and chewier the candy. A basic formulation for marshmallows includes 1–2.5% gum arabic, 0.35% gelatin, 37% sugar, 19.3% dextrose, 1.8% albumen, 0.35% salt, 0.35–1.0% vanilla, 23.4% corn syrup, 42 dextrose equivalents and hot water to 100%. Production includes blending, addition of corn syrup and water, mixing, adding vanilla when thickening has occurred, molding and drying (Davidson, 1980).

9.8 Fluid and emulsion gels

Modifying the material properties of gums can impart a fat/emulsion-like behavior. Limited success has been achieved by manufacturing de-watered fiber-based substitutes, carbohydrate-based particulates and protein precipitates for optimal mouth performance. An alternative is producing gel particulates by forming gel in shear, or breaking the gel into particles by shearing or chopping. These products are called fluid gels and they can be successfully produced from many gelling agents, such as carrageenans, agars, agarose, alginates, pectins and gellans, among others (Norton *et al.*, 1998). The properties of the fluid gels depend upon the gum used for the preparation and its particle size. Such products are a consequence of aggregate formation by hydrocolloids that have undergone de-watering and are trying to precipitate from solution. More information regarding these special moieties, creation mechanisms, stability and the effect of temperature on them can be found elsewhere (Norton *et al.*, 1998, 1999, 2000). Gellan gum is a multi-functional hydrocolloid for use in a wide variety of food products as a gelling-texturizing, stabilizing, film-forming, suspending and structuring agent. It can also be used, like other gelling agents, to prepare fluid gels. Fluid gel formation is commonly employed in the production of custards and gravies, the fluid mass formed during homogenization or high temperature short time/ultra-high temperature (HTST/UHT) processing (Sworn *et al.*, 1995).

An emulsion gel is a moiety that contains oil droplets and can serve as a basis for future foods as well as being a good food model. The effect of oil droplets on the rheological properties and micro-structure of mono-dispersed oil-in-water emulsion agar gels prepared by membrane-emulsification technique was investigated by mechanical measurements and scanning electron microscopy (SEM) observation. The compressive gel strength of the emulsion gel decreased with an increase in oil droplet size, while that of the emulsion gel determined by puncture test showed almost no change. Cryo-SEM observation revealed that oil droplets aggregate in the emulsion gel and that the gel has some void spaces between the gel network and the oil droplet aggregate (Gohtani *et al.*, 2000).

9.9 References

Aguilera, J.M. (1992) Generation of engineered structures in gels. In: (Schwartzberg, H.G. & R.W. Hartel, Eds.), *Physical Chemistry of Foods*. Marcel Dekker Inc., NY, Basel and Hong Kong.

Aguilera, J.M. & D.W. Stanley (1986) In: (Le Maguer, M. & P. Jelen, Eds.), *Food Engineering and Process Applications*, vol. 2. Elsevier Applied Science Publishers, London, 131.

Ben-Zion, O. & A. Nussinovitch (1996) Predicting the deformability modulus of multi-layered texturized fruit and gels. *Lebensm.-Wiss. u.-Technol.*, **29**, 129–134.

Davidson, R.L. (1980) *Handbook of Water-soluble Gums and Resins*. McGraw-Hill, NY.

Fellows, P.J. (2000) *Food Processing Technology: Principles and Practice*. CRC Press, Boca Raton, NY & Woodhead Publishing Limited, Cambridge, England.

Glicksman, M. (1969) *Gum Technology in the Food and Other Industries*. Academic Press, NY.

Glicksman, M. (1976) Fabricated food. *Cereal Foods World*, **21**, 17–26.

Gohtani, S., K.H. Kim & Y. Yamano (2000) Rheological properties and microstructure of monodispersed O/W emulsion agar gel. In: (Nishinari, K., Ed.), *Hydrocolloids*, part 1. Elsevier, Amsterdam, The Netherlands, pp. 91–96.

Howes, F.N. (1949) *Vegetable Gums and Resins*. Chronica Botanica, Waltham, MA.

Kaletunc, G., A. Nussinovitch & M. Peleg (1989) Alginate texturization of highly acid fruit pulp and juices. *J. Food Sci.*, **55**(6), 1759–1761.

Kong, J.Y., T. Yano, J.D. Kim, S.K. Bae, M.Y. Kim & I.S. Kong (1994) Prediction of effective thermal-diffusivity of fish and meats. *Biosci. Biotechnol. Biochem.*, **58**(11), 1942–1946.

Lillford, P.J. (1985) In: (Simatos, D. & J.L. Multon, Eds.), *Properties of Water in Foods*. Nijhoff Publisher, Dordrecht, p. 543.

Lodge, N. (1981) Kiwi fruit: two novel processed products. *Food Technology in New Zealand*, **16**(7), 35, 37–38, 41.

Lugay, J.C. & M.K. Kim (1981) In: (Stanley, D.W., E.D. Murray & D.H. Lees), *Utilization of Protein Resources*. Food and Nutrition Press, Westport, CT, p. 177.

Luh, N., M. Karel & J.M. Flink (1976) A simulated fruit gel suitable for freeze dehydration. *J. Food Sci.*, **41**, 89–93.

Matz, S.A. (1984) *Snack Food Technology*. AVI Publishing Co., Westport, CT.

Mitchell, J.R. & J.M.V. Blanshard (1976) Rheological properties of alginate gels. *J. Texture Studies*, **7**, 219–234.

Norton, I.T., T. Foster & R. Brown (1998) The science and technology of fluid gels. In: (Williams, P.A. & G.O. Phillips, Eds.), *Gums and Stabilisers for the Food Industry*, vol. 9. The Royal Society of Chemistry, Cambridge, UK, pp. 259–268.

Norton, I.T., D.A. Jarvis & T.J. Foster (1999) A molecular model for the formation and properties of fluid gels. *International J. Biological Macromolecules*, **26**(4), 255–261.

Norton, I.T., C.G. Smith, W.J. Frith & T.J. Foster (2000) The production, properties and utilization of fluid gels. In: (Nishinari, K., Ed.), *Hydrocolloids*, part 2. Elsevier, Amsterdam, The Netherlands, pp. 219–227.

Nussinovitch, A. (1993) Gum-based texturized products. In: *Yearbook of Science and Technology*. McGraw-Hill, NY, pp. 138–140.

Nussinovitch, A. (1997) *Hydrocolloid Applications Gum Technology in the Food and Other Industries*. Blackie Academic & Professional, London, UK.

Nussinovitch, A. & M. Peleg (1990) Mechanical properties of a raspberry product texturized with alginate. *J. Food Process. Preserv.*, **14**, 267–278.

Nussinovitch, A., M. Peleg & M.D. Normand (1989) A modified Maxwell and a nonexponential model for characterization of the stress relaxation of agar and alginate. *J. Food Sci.*, **54**, 1013–1016.

Nussinovitch, A., I.J. Kopelman & S. Mizrahi (1991a) Mechanical properties of composite fruit products based on hydrocolloid gel, fruit pulp and sugar. *Lebensm.-Wiss. u.-Technol.*, **24**, 214–217.

Nussinovitch, A., I.J. Kopelman & S. Mizrahi (1991b) Modeling of the combined effect of fruit pulp, sugar and gum on some mechanical parameters of agar and alginate gels. *Lebensm.-Wiss. u.-Technol.*, **24**, 513–517.

Oakenfull, D., K. Nishinari & E. Miyoshi (2000) A comparative study of milk formed with κ-carrageenan or low-methoxy pectin. In: (Nishinari, K., Ed.), *Hydrocolloids*, part 2. Elsevier, Amsterdam, The Netherlands, pp. 153–163.

Ochi, A., K. Katsuta, E. Maruyama, M. Kubo & T. Ueda (2000) Effects of sugars on stability of egg foam and their rheological properties. In: (Nishinari, K., Ed.), *Hydrocolloids*, part 2. Elsevier, Amsterdam, The Netherlands, pp. 275–280.

Philp, K., Z. DeFretius, D. Nicholson & R. Hoffman (1998) Tiger striping in injected poultry its causes and cures. In: (Williams, P.A. & G.O. Phillips, Eds.), *Gums and Stabilisers for the Food Industry*, vol. 9. The Royal Society of Chemistry, Cambridge, UK, pp. 276–285.

Rasanen, J. & K. Autio (1998) Effect of stabilisers and water content on the rheological properties of prefermented frozen wheat doughs. In: (Williams, P.A. & G.O. Phillips, Eds.), *Gums and Stabilisers for the Food Industry*, vol. 9. The Royal Society of Chemistry, Cambridge, UK, 286–294.

Ring, S.G. & G. Stainsby (1982) In: (Phillips, G.O., D.J. Wedlock & P.A. Williams, Eds.), *Progress in Food and Nutrition Science*, vol. 6, *Gums and Stabilisers for the Food Industry: Interactions of Hydrocolloids*. Pergamon Press, Oxford, pp. 323–329.

Silberberg, A. (1989) In: (Glass, J.E., Ed.), *Polymers in Aqueous Media Performance Through Association*. Advances in Chemistry Series No. 223, American Chemical Society, Washington, DC, p. 3.

Sworn, G., G.R. Sanderson & W. Gibson (1995) Gellan gum fluid gels. *Food Hydrocolloids*, **9**(4), 265–271.

Szczesniak, A. (1968) Simulated fruits and vegetables. United States Patent 3,362,831.

Tolstoguzov, V.B. (1971) Method of preparing artificial foodstuffs. USSR Patent 296,554.

Tolstoguzov, V.B. & E.E. Braudo (1983) Fabricated foodstuffs as multicomponent gels. *J. Texture Studies*, **14**, 183–212.

Topalian, H. & C.C. Elsesser (1966) Table syrups. United States Patent 3,282,707.

Torres, A. (1977) *Textural properties of amylose and starch sponges*. M.Sc. thesis, Department of Food and Agricultural Engineering, University of Massachusetts, Amherst, MA.

Uniliver Ltd. (1974) Simulated fruit. British Patent 1,369,198.

Weiner, G. & A. Nussinovitch (1994) Succulent, hydrocolloid-based, texturized grapefruit products. *Lebensm.-Wiss. u.-Technol.*, **27**, 394–399.

Werbin, S.J. (1960) Practical aspects of viscosities of natural gums. In: *Physical Function of Hydrocolloids*. American Chemical Society, Washington, DC, pp. 5–10.

Wood, F.W. (1975) Artificial fruit and process therefore. United States Patent 3,892,870.

Abbreviations

ADI	acceptable daily intake
AFM	atomic force microscopy
AG	arabino-galactan
AGP	AG–protein complex
AGs	acetylated monoglycerides
ASTM	American Society for Testing and Materials
BFBR	bi-particle fluidized-bed reactor
BW	bees wax
CA	controlled atmosphere
CFU	colony forming units
CMC	carboxymethyl cellulose
COD	chemical oxygen demand
DD	de-acetylation
DE	dextrose equivalent
DM	degree of methoxylation
DP	degree of polymerization
DSC	differential scanning calorimetry
E	elongation at break
EP	ethylene permeability
FDA	food and drug administration
FOS	fructose oligosaccharides
GDL	glucono-δ-lactone
GI	glycoprotein
GRAS	generally recognized as safe
HEC	hydroxyethylcellulose
HMP	high-methoxy pectin
HOPG	highly oriented pyrrolite graphite
HPC	hydroxypropyl cellulose
HPMC	hydroxypropylmethyl cellulose
HTST	high temperature short time
ICP	inductively coupled plasma
KM	konjak mannan

LBG	locust bean gum
LMP	low-methoxy pectin
MA	modified atmosphere
MC	methyl cellulose
MCC	microcrystalline cellulose
OP	oxygen permeability
OURm	maximum oxygen uptake rate
PEG	polyethylene glycol
PEO	polyethylene oxide
PGA	propylene glycol alginate
PVA	polyvinyl alcohol
PVP	polyvinyl pyrrolidone
RH	relative humidity
RPBR	recycling packed-bed reactor
SEM	scanning electron microscope
SHMP	sodium hexametaphosphate
SPI	soy protein isolate
SSM	skin surface model
STM	scanning tunneling microscopy
TTI	time–temperature integrator
UF	ultra filtration
UTM	universal testing machine
VAD	vitamin A deficiency
WVP	water vapor permeability

Index